Biology

Morton Jenkins

Biology

Morton Jenkins

TEACH YOURSELF BOOKS

Acknowledgements

The author expresses his sincere thanks to Helen Green of Hodder & Stoughton for generating the idea of the book and for her help and encouragement throughout its production. It is also a pleasure to thank Dr Sue Noake, Headteacher of Lewis Girls Comprehensive School, Ystrad Mynach for her advice and help during the preparation of the text.

For UK order queries: please contact Bookpoint Ltd, 130 Milton Park, Abingdon, Oxon OX14 48B. Telephone: (44) 01235 827720. Fax: (44) 01235 400454. Lines are open from 9.00–18.00, Monday to Saturday, with a 24-hour message answering service. Email address: orders@bookpoint.co.uk

For U.S.A. order queries: please contact McGraw-Hill Customer Services, P.O. Box 545, Blacklick, OH 43004-0545, U.S.A. Telephone: 1-800-722-4726. Fax: 1-614-755-5645.

For Canada order queries: please contact McGraw-Hill Ryerson Ltd., 300 Water St, Whitby, Ontario L1N 9B6, Canada. Telephone: 905 430 5000. Fax: 905 430 5020.

Long renowned as the authoritative source for self-guided learning – with more than 30 million copies sold worldwide – the *Teach Yourself* series includes over 300 titles in the fields of languages, crafts, hobbies, business and education.

British Library Cataloguing in Publication Data
A catalogue record for this title is available from The British Library.

Library of Congress Catalog Card Number: On file

First published in UK 2001 by Hodder Headline Plc, 338 Euston Road, London, NW1 3BH.

First published in US 2001 by Contemporary Books, A Division of The McGraw-Hill Companies, 4255 West Touhy Avenue, Lincolnwood (Chicago), Illinois 60712–1975 U.S.A.

The 'Teach Yourself' name and logo are registered trade marks of Hodder & Stoughton Ltd.

Typeset by Transet Limited, Coventry, England.
Printed in Great Britain for Hodder & Stoughton Educational, a division of Hodder Headline Plc, 338 Euston Road, London NW1 3BH by Cox & Wyman Ltd, Reading, Berkshire.

Impression number   10 9 8 7 6 5 4 3 2 1
Year   2007 2006 2005 2004 2003 2002 2001

CONTENTS

INTRODUCTION

Stones grow, plants grow and live, animals grow, live and feel.

Carolus Linnaeus (1707–1778)

Has the way the basics of biology are taught changed since Linnaeus made this statement in the eighteenth century? We have come a long way since then and perhaps a casualty of specialisation is the failure to appreciate the wonder of a whole subject. If we concentrate on a single bloom, the magnificence of the whole bouquet is lost. Today, we live in a world of specialists. No one can really be a 'biologist', in the same way as scientists can no longer be omniscient. Even the specialisms of biology become sub-divided; for example geneticists specialise in subjects like gene cloning or medical genealogy and cytologists specialise in cell membranes or in other specific parts of a cell. To the non-specialist, non-scientist, the word 'biology' still has meaning as the study of living things. Many may be reminded of their school days when, all too often, the subject was reduced to necrology – an obituary of sloppy, grey, dead things in jars of liquid – things which were no more than laundry lists of jaw-breaking names which had to be remembered for homework for a test the next day. The rigidity of school curricula, in order to master certain assessment criteria, often snuffs out the initial flame of wonder and interest that many youngsters have in the living world. Perhaps too much emphasis is placed on formal assessment. An analogy is a legend of a farmer who spent so much time weighing his pig that he did not have time to feed it properly and so the pig died! Are we in danger of spending so much time assessing, that we do not have time to teach properly? Those topics which are easiest to assess are not necessarily the most interesting to learn, and those topics which are interesting to learn are often very difficult to assess objectively. How do you assess appreciation of a topic rather than knowledge of it? Often, people of an older generation remember their school biology as little more than 'frogs, ferns and fornication' or 'cutting up earthworms and eyes', with little understanding of the fundamental principles which unify life on Earth. The blunt truth is that school text books are written to be sold and they won't be sold unless they meet the criteria of official specifications for awarding authorities. Hence, much of the excitement and fascination found in children regarding living things, often becomes stifled when schools have to prepare them for assessments using standard texts. This book is an attempt to move away from the plethora of text books on biology. It is meant to concentrate on those aspects of the subject which need not necessarily be assessed formally. Hopefully, it will give the reader a background to the subject as viewed from a height, highlighting the extraordinary nature of biology which makes it different to the other sciences.

On Earth there are more than a million kinds of living things. They are found on land, in water and in the air. Some are mere blobs, ugly to look at and hard to recognise as being alive. Others are intricately formed, beautifully coloured and elegant in motion. In size they range from tiny specks, visible only with the highest powered microscopes, to whales and giant trees. This book is about such living things interacting with their rich environment. The use of technical terms has been minimised but, inevitably, biological terminology must occur in some of the descriptive prose used to communicate concepts. In order to help the reader, there is a glossary at the end of the book and a short guide to 'biospeak' which attempts to give the Greek and Latin derivations of some of the more common words used in biology. Since the invention of its name in 1802 by Jean Baptiste Lamarck, biology has, of necessity, grown into a tree of specialisms. Each specialism is at the tip of a branch and each worthy of a *Teach Yourself* book. Perhaps, after having a taste of some of the main topics of biology the reader will be encouraged to explore further. If this is the case, the book will have achieved its aim. See Further Reading for specialist biology books in the *Teach Yourself* series.

1 | LIFE AS WE KNOW IT

Solitary confinement

In contrast to things that have never lived, living things can take non-living material from outside their bodies and transform it into themselves, each according to its own pattern. Thus a child and a kitten can eat the same cooked meat. In one case the dead meat is turned into human, and in the other, into cat. This transformation requires a special type of chemical called deoxyribonucleic acid or DNA (see p.101). So, if we really have an uncontrollable urge to define life, we can say it is a self-sustaining system dependent on DNA. However, rather than philosophising over the meaning of life, let us begin a study of biology by looking at the ways in which living things are put together, what they contain, and how their chemistry makes them tick. These pursuits could be an endless quest because simple answers to simple questions often lead to more complex questions with even more complex answers. The end product is a biosphere of unimaginable complexity and beauty. Knowing even a little about life cannot fail to amaze us that it happens at all.

Make a list of the parts of some living things: skin, bone, muscle, liver, roots, stems, flowers etc. If a thin slice of any one of these is examined with a microscope it will be seen to be composed of apparently isolated tiny compartments – ubiquitous sacs of life – called cells. Although millions of microscopic examinations have confirmed this observation, it all began with the curiosity of Robert Hooke (1635–1703). He published his meticulously detailed descriptions of cells in one of the most significant science books ever written. This was his *Micrographia* of 1665. His most widely known observations took place with a quite unlikely material – a slice of cork; but with this, the science of cytology (the study of cells) was born.

What he saw reminded him of the small rooms in which monks lived, the cells of a monastery. In fact he was observing an interlocking arrangement of plant cell walls. All of their contents had long since died in boxed solitary confinement. The empty boxes that he saw had once held active living materials. There was no suggestion in Hooke's report that his discovery would be of great biological importance – he did not realise that he had seen the building blocks of life itself.

The Royal Society of London published Hooke's masterpiece, and because King Charles II was the patron of the Society, Hooke began his book with a dedication: 'To the King, Sir, I do here most humbly lay this small present at Your Majesty's Royal feet'. It may have seemed a small present for a king, but it became a priceless gift to Mankind.

Unknowingly, then, Hooke had laid the foundations for a new science but about a century was to pass before the significance of his work was grasped. It was not until 1805 that the German naturalist, Lorenz Oken, formally published what has become known as the cell theory: 'All life comes from cells and is made of cells'. Oken's work was confirmed in 1839 by two other Germans, botanist Matthias Jakob Schleiden and zoologist Theodor Schwann, who independently concluded that all living things are made of cells. Today, we know that the position of viruses in this definition is questionable, but by saying 'living things' we mean things that can reproduce independently. Viruses are exceptional because they are all parasitic and can only reproduce inside other living things.

The work of Schleiden and Schwann promoted the cell theory which is probably the most important biological generalisation of the first half of the nineteenth century. The cell theory is to biology as the atomic theory is to physics and chemistry. It has grown in importance and is central to the continued development of modern biology. After publication of the cell theory, Schwann became a professor at the University of Louvain and, after nine years, moved to the University of Liege. He was known as an outstanding experimenter and an excellent teacher. Schleiden was a successful lawyer whose interest in science was so compelling that he gave up his law practice and, after graduating in medicine, devoted himself mainly to plant science. It was Schleiden who encouraged Schwann to finally develop and publish the cell theory in a research paper.

So here was an unlikely duo of zoologist and botanist. Schwann was gentle and kind, always avoiding controversy, whereas Schleiden was assured, disputative and certain to provoke discussion. They complemented each other in this single joint effort and provided an exemplary model of co-operative research. Schleiden published first in 1838 and reached two major conclusions. He stated that plants were built up of cells, and that the embryo of a plant arose from a single cell. Schwann carried out the more comprehensive work and first used the term, cell theory. He published this theory in 1847 under the title *Microscopical Researches into the Accordance in the Structure and Growth of Animals and Plants*.

In 1860, the German pathologist, Rudolf Virchow asserted, in a succinct Latin phrase, that all cells arise from cells. He showed that the cells in diseased tissue had been produced by the division of originally, normal cells. By the end of the nineteenth century many observations of the microscopical structure of cells had been published with the aid of microscopes which relied on the transmission of light. At that time, microscopes had been developed which had reached the limits of magnification possible with lens systems. However, the question remained: are cells made only by chemicals, or do they contain some special ingredient that escapes detection, a mysterious force that makes life work? The dispute over the answer to this has been one of the longest and hottest of all quarrels in the scientific world. In order to address this problem, a new invention – the transmitting electron microscope, became invaluable.

A carpenter could not make a living without tools of the trade – perhaps the same design that has not changed since the time of classic medieval master builders. Hundreds of years ago, carpenters used their muscles to power their tools just as they do in the twenty-first century. The invention of power tools changed this and although the end products depend on the carpenter's skill, the method of powering his tools is different. As an analogy, the pioneering cytologist's tool kit included the ordinary light microscope which was once the only way to obtain the degree of magnification needed to view cells. There have been many variations on its theme, but the basic principle has remained the same, but so have the limitations of the technique – limitations imposed by the light microscope's power of resolution.

Resolution refers to the smallest distance by which two points can be separated and still be distinguished as separate points. If they are separated by a distance smaller than the resolving power of the microscope, they will be seen as a single point. Under ideal conditions, the best light microscopes have a resolution of about 0.25 micrometres with ordinary white light. The power has been improved with the use of ultraviolet light where the UV light is projected on to a specimen which absorbs the light energy and re-emits it as electrons that are then magnetically focused. This is the photoelectron microscope.

The resolving power of the transmitting electron microscope (TEM) is of an order of magnitude greater than that of a light microscope. Thus cells can be examined in far greater detail. The TEM was invented in the 1930s and works on an entirely different principle from the light microscope; just as the power motor works on an entirely different principle to muscles in our carpenter's analogy. In the TEM, excited electrons are drawn from a heated filament and are then directed, or focused, by magnetic fields. Theoretically a resolution of about 0.005 nanometre is possible.

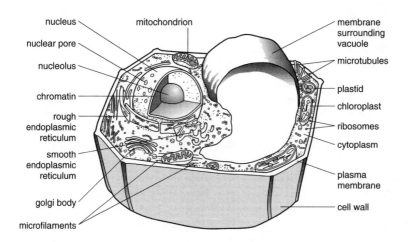

Figure 1.1   Plant cell

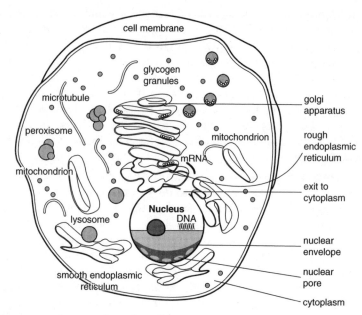

cell membrane
glycogen granules
microtubule
peroxisome
mitochondrion
mitochondrion
mRNA
lysosome
Nucleus
DNA
smooth endoplasmic reticulum
golgi apparatus
rough endoplasmic reticulum
exit to cytoplasm
nuclear envelope
nuclear pore
cytoplasm

Figure 1.2   Animal cell

With this tool, cytologists have mapped the cell's surface and penetrated below it to the interior. The level of detail to be seen is shown in figure 1.2.

Our cells are like living factories with the nucleus representing the management, the other organelles the machinery, and the enzymes (see p.28) the workers. Just like a factory, the cell must obtain raw materials for it to perform efficiently. It must also be able to get rid of waste. In a cell, this exchange is controlled by the plasma membrane (cell membrane), which regulates the entry and exit of materials like a security system at a factory's gates. These 'gates' are composed of lipids (fats) and proteins in combinations with other chemical compounds (see fig. 1.3)

Whether or not materials pass through the gates depends on the arrangement of lipid and protein molecules within the membrane. Some protein molecules sit on its surface while others penetrate it. They can help to carry substances across the membrane by combining with them. By escorting the materials through in this way they facilitate the process of diffusion (see p.29) by which substances pass from where they are in high concentration to where they are in low concentration. Some molecules are so small that they simply diffuse across without any help. These include oxygen, carbon dioxide, and water. In some cases the cell needs to accumulate a particular substance against a concentration gradient – i.e. the substance has to move from where it is in low concentration to where it is in high concentration. For this to take place, a process of active transport is needed which requires the factory workers (the enzymes) to release energy to be used in pumping materials across the membrane.

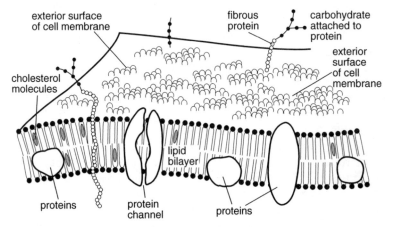

Figure 1.3   Structure of a cell membrane

A further means of intake of materials involves a process called phagocytosis (from the ancient Greek, (*phage* = eating; *cyton* = cell). So it means cells that 'eat' things. In this process, the membrane flows around the 'food' being taken in, forming a bubble-like vesicle which becomes detached on the inside of the cell membrane. Once inside the vesicle, digestion begins using enzymes made by the cell. A good example of this is when certain white blood cells digest bacteria which may enter our bodies.

Besides being essential as selective barriers to the external environment of cells, membranes are also important within the cells. They divide the factory into compartments, where different functions are carried out by different workers – the enzymes. A network of membrane-bound tubes forms the endoplasmic reticulum (from the ancient Greek, *endo* = inside; *plasm* = moulded form; *reticulum* = network). In the cell it acts like an internal transport system – like a factory's intranet system of communication. It also presents a large surface area for the attachment of structures called ribosomes, where proteins are assembled on a sort of conveyor belt system in our factory analogy.

Then, there has to be a packaging department. It was first discovered as long ago as 1898 when Camillo Golgi (1844–1926), an Italian physician and Nobel Prize winner, was experimenting with some cell-staining techniques. He discovered that when he stained cells with silver nitrate, strange shapes appeared in the cells. However, it took half a century before the significance of this discovery was appreciated because when Golgi first described his 'Golgi bodies', other microscopists thought that they were merely artifacts or debris produced by the staining process because they could not be seen in *living* cells. The function of the Golgi body was not discovered until well after the invention of the electron microscope. It is now known that its role is complex but, briefly, it is where many substances made by the cells are stored and sometimes modified before being packaged into membrane-bound sacs and transported to the outside of the cell. It is rather like a packer for an internet home delivery service.

All factories need a source of power. Our cells have tiny membrane-bound compartments called mitochondria (from the ancient Greek *mito* = thread and *chondro* = grain-like). The naming of them thus describes their appearance when stained and seen with the highest magnification of a light microscope. The electron microscope has elucidated the details of their structure. They appear in a variety of shapes – round, elongate and thread-like. In some cases they seem to wriggle around the cell, while in others they seem to be immobile. They are not randomly distributed, but tend to be aggregated in places where most energy needs to be released. They are the power-houses of the cell where most aerobic respiration takes place (see p.33).

All of these tiny organelles that we are comparing to parts of a factory are embedded in the cytoplasm which is about 90% water and a mixture of mineral salts and giant molecules. These react together as the living contents of all cells. When it contains a nucleus it is called protoplasm or 'first moulded form', a name given to it by the Czech physiologist, Jan Evangelista Purkinje, in 1839. The term was extended to mean the contents of all cells by the German botanist, Hugo von Mohl.

Our building kit

How many building blocks does it take to make one of us? How big are our building blocks and are they all the same?

On average we can say that most cells, at least the ones in our bodies, measure about 1000 to a centimetre. There are about 50 million, million cells in the body of an average human and about 2000 times as many in a whale. The giants of the cell world always come from females. These are their eggs with the largest being that of the ostrich. Besides birds' eggs, other large cells are the eggs of fish and frogs. Males on the other hand, produce some of the smallest of cells but make them in much larger numbers. These are the spermatozoa (from the ancient Greek, *sperm* = seed; *zoo* = animal).

Cells vary not only in size but also in the job they perform. Liver cells carry out primarily chemical tasks related to dealing with digested food and clearing the blood of poisonous chemicals. It is so important that it is made of thousands of millions of cells.

A hen's egg or fish egg has quite a different job – to protect and nourish the baby chick or fish until it is ready to hatch. The food and water stored in them contribute to their size. Before birth, human babies and those of most other mammals are nourished by their mother's blood stream. Thus mammalian eggs are comparatively small on the egg scale of things but large on the cell scale. Sperm cells, on the other hand, are all small. Their only job is to provide a nucleus to fuse with the egg nucleus to make the fertilized egg called a zygote (from the ancient Greek *zygo* = joined), from which the embryo develops by a process of cell division (see p.22). So that the sperm cells can reach the egg, either in water or inside the female's body, each has a filament-like 'tail' which helps it in swimming to its target. Most cells specialise to carry out certain jobs. A nerve cell carries messages in the form of electrical impulses; others contract and relax to make bits of us move; a red blood cell carries oxygen combined with haemoglobin; certain white cells gobble up invading bacteria and viruses or make chemicals to protect us from disease. Such specialised cells make up a tissue, and organs are made up of groups of tissues.

The specialised cells are shaped to carry out their jobs. Muscle cells are shaped like rods or long spindles, whose tapering ends fit neatly together as a sheet of tissue. Nerve cells have tree-like branches, one of which may be particularly long so that it can carry electrical impulses to the next to form a chain of interconnected cells functioning as a single unit. This is an ideal arrangement for getting messages to and from the brain – rather like a system of insulated, conducting wires in an electric circuit. Cells lining a moist surface, such as the inside of the windpipe and lungs, may have tiny hair-like projections called cilia. They stick out from the surface to pass particle-containing mucus along as they wave like a field of wheat in the wind. Among other types of cells are gland cells that make and release special fluids called secretions; connective tissue cells that fill spaces between tissues; and epithelial, or skin cells, that cover surfaces for protection.

Gland cells are often jug-shaped, while cells for covering large areas are rather flattened. Similarly, cells covering the outside of leaves and stems are also flat. Inside leaves, the most prominent cells look rather like squat columns and many of the cells in plant stems are long tubes, a shape which makes them ideal for carrying solutions to parts which need them.

All in one

Some cells live as independent organisms and can do all of the things that multi-cellular organisms can do. Among the larger one-celled creatures is *Amoeba*. It is roughly four times as big as a human egg. When we get down to bacteria, however, size diminishes rapidly. These tiny microbes are just about the smallest living things that can be seen with a light microscope and have many features which are considered to be more primitive than our cells. For example, they do not have a membrane around their

DNA, so therefore they cannot be said to have a distinct nucleus. Because of this, they are placed in a group of living things called the prokaryotes (from the ancient Greek *pro* = before; *kary* = nut/ nucleus), as opposed to all cells with a proper nucleus, the eukaryotes (*eu* = proper; *karyo* = nucleus), (see p.8).

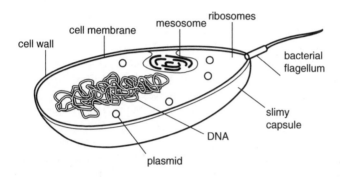

Figure 1.4   Prokaryote

Prokaryotes also do not have mitochondria, which leads us to an interesting idea regarding the origin of both mitochondria and chloroplasts (see p.7). There is now much evidence to suggest that mitochondria and chloroplasts may be the descendants of free-living prokaryotic cells that invaded other cells during life's evolutionary history. This idea is known as the symbiosis hypothesis. Symbiosis is defined as a relationship between two organisms of different species in which both gain benefit. The idea goes back to the 1970s when the American cytologist, Lynn Margulis, suggested that mitochondria and chloroplasts arose from bacterial and blue-green algal ancestors. It is possible that they were taken in by some of the earliest developed eukaryotes and that they became mutually dependent on their host cells and unable to exist outside them. In time, the co-habitants became permanent partners, unable to separate. Mitochondria and chloroplasts can reproduce themselves, but in modern-day eukaryotes, their reproduction is often precisely synchronised with the reproduction of the host cell. They are also surrounded by a double membrane, similar to cell membranes of eukaryotes. So, are bacteria the smallest living things? Many diseases of both plants and animals are caused by viruses. These cannot be considered as being made of cells because they do not have any structures that we associate with cells and can only reproduce while parasitising host cells.

Is any one part of a cell more important than another? One way to find the answer to this question is to look closely at some of the independently living cells, such as the tiny animal, *Amoeba*. This is probably the most commonly remembered of all the Protozoa (single-celled animals) by people with a vestigial knowledge of school biology. The structure and behaviour of this tiny animal has been taught to students since the nineteenth century – in most cases without ever having been seen by the teacher or pupil! The most talked about *Amoeba* is *Amoeba proteus*, a fresh water species. There are many other types, some living in the sea, others living in the intestine, causing amoebic dysentery, and yet others wandering around between the teeth of your pet dog or cat!

However, *Amoeba proteus* is the one that we were taught about as a minute blob that could move by squeezing out projections of protoplasm on the bottom of ponds. It was discovered as long ago as 1755 by Rozel von Rosenhof. He named it *Proteus animalicule* because it can change its shape so often and assumes so many varieties of shape. In Greek mythology, Proteus was an 'old man of the sea', who looked after the seal flocks of Neptune. He had a great reputation as a prophet and possessed the gift of endless transformation, adopting all manner of shapes and disguises in order to escape from those enquiring people who wished him to make prophesies for them. The name *Proteus animaliculae* was changed to *Amoeba* from the Greek meaning 'change'. Professor T.H. Huxley, in the nineteenth century (best known for supporting Charles Darwin's theory of evolution) described it as 'one of our most wonderful animals, for it walks without legs, eats without a mouth, and digests without a stomach!'

In spite of the fact that amoebae are just about as simple as you can be and still be called an animal, they seem to get along very well with very little in the way of internal structure to clutter up their protoplasm. Indeed, if you think about it, because they only reproduce asexually by splitting into two, the ones crawling around today must be almost identical to their ancestors – perhaps hundreds of millions of years ago!

Although it has neither brain nor sense organs, an amoeba can move toward food and away from conditions that might harm it.

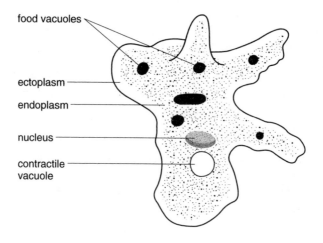

food vacuoles

ectoplasm

endoplasm

nucleus

contractile vacuole

Figure 1.5   *Amoeba*

It does this by extending part of its body and then flowing into it – a bit like squeezing toothpaste out of a toothpaste tube. The extensions are called pseudopoda (from the ancient Greek *pseudo* = false; *pod* = food). When it reaches the bacteria or small plants (algae) upon which it feeds, it simply flows around its meal. Thus the food is inside the amoeba in a bubble called a vacuole, surrounded by cytoplasm. In time, the contents of the vacuole will be digested and the products will be absorbed into the cytoplasm. Any remaining indigestible bits, like the cell wall of plant cells, will be eliminated by a simple procedure of working the vacuole to the surface of the amoeba and simply flowing away from it. Excess water and other waste products of the amoeba's metabolism are emptied into a contractile vacuole, which periodically bursts to excrete the waste.

Size matters

As always happens with living things, a well-fed *Amoeba* grows. Now there is a problem. At least one of the 'B' movies of the 1950s depicted a giant blob of protoplasm, about the size of a large bus, oozing around the countryside engulfing anything slower than itself. In reality, however, the largest single-celled animal you are likely to see will be with your microscope and will be no bigger than a pinhead – and that will be a giant among the Protozoa! Besides the fact that all protozoans live in water, and that a blob of protoplasm, the size of a bus, would be too heavy to move without a skeleton, the main problem is the ratio of volume to surface area.

As cells grow, the volume of their protoplasm increases more than the area of their surface membranes. Because *Amoeba* does not have a fixed shape, this is hard to visualise; but consider a 1cm cube. Its volume is 1cm^3 and its surface area is 6cm^2. Double the edge of the cube, and the volume becomes 8cm^3, while the surface area becomes 24cm^2. Double the edge once more and the volume will jump up 8 times to 64cm^3 while the surface area increases to

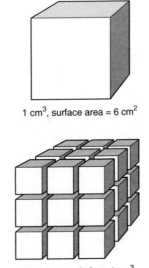

1 cm^3, surface area = 6 cm^2

27 cubes made from 1 cm^3,
surface area = 17.8 cm^2

Figure 1.6   How to increase surface area using the same volume

$96cm^2$ – 4 times what it was. An amoeba, of course is not a cube, but the principle holds. As the volume of cytoplasm increases by a factor of eight, the amoeba's surface area increases by a factor of four. This is true of all cells. If growth continued, the inside of the cell would soon be so great that the membrane could not get essential oxygen in fast enough for the cell's needs; neither could harmful waste carbon dioxide leave fast enough.

In *Amoeba*, when it is time to divide, the nucleus splits into two parts which move a little distance from each other. Then the cytoplasm pinches down between the two nuclei, and the cell divides. The new cells, often called daughter cells, move off to feed, grow and divide again. From this it would seem almost as if the cytoplasm and membrane control cell activities – but experiments have told a different story.

With a good microscope, a steady hand, a fine glass needle, and endless patience, scientists have been able to perform microsurgery on amoebas, cutting off parts or removing the nucleus. In some experiments, almost all the cytoplasm has been lopped off. Although the cut-off portion soon dies, the surviving fragment, containing a portion of the cytoplasm and the nucleus, continues to feed and grow. In one experiment a piece of cytoplasm was removed every day for four months and the amoeba never divided. A control amoeba, that was not operated upon, went through 64 divisions during the same period. In other experiments, the nucleus was removed. When this was done, the amoeba continued to move around. The contractile vacuole formed and emptied, indicating that water, at least, passed into the cell and that some metabolism was maintained. These amoebas, without a nucleus, did not feed and did not divide.

Now we come back to our original question. Is one part of a cell more important than the others? The answer is 'no'. Although the nucleus directs and controls all activities, the cytoplasm is the place where feeding, movement and growth occur. A complete cell is capable of all life's functions. A piece of a cell is a fragment of life without a future.

The miracle of growth

A newborn infant has thousands of millions of cells. Where do they come from? As a rule, most living things that are made of many cells begin life as a single cell, the fertilised egg, that results when a male sex cell fuses with a female sex cell. During the period before birth, the cell divides again and again. As division goes on, the cells become specialised into tissues and organs. The new life form becomes transmuted from a shapeless lump of cells into a recognisable organism.

Cell division does not cease at the time of birth. The miracle of life is accompanied by growth. When cells in an organism divide, the new cells are smaller than the original cells were, but gradually they increase in size and get ready to divide again; and so the process continues. Some cells die, such as those in the outer layers of your skin or the outer layers of tree bark. These cells have long since given up the ability to divide. Other cells are 'asleep', or dormant, such as those in a seed; they do not divide until are 'awakened' by a change in environmental conditions. On the other hand, well-fed bacteria may divide every twenty minutes. If all the descendants of one bacterium survived, they would become some five thousand million in twelve hours! A maize stalk may grow several centimetres in a day. A healthy human baby may put on weight at the rate of several grammes per day – a milligram every three seconds. A cell weighs somewhere between a thousandth and a millionth of a milligram, so that the baby has acquired between a thousand to a million new cells in those three seconds. However, this very active growth is only a transient stage. Growth stops in the maize stalk and begins in the seeds as they begin to form. A baby grows more and more slowly until it reaches the teenage adult stature.

Although growth in height ceases in humans, some cells keep their power to divide. When you cut your finger, you damage and kill some living cells. Within a few hours, the surviving cells around the wound begin to grow and divide, and within a few days the dead cells are replaced with new tissue. If you break a bone, new bone tissue grows to join the pieces together, and every second of the whole of your life, your red bone marrow is producing millions of blood cells to replace those which become used and worn out.

Primitive animals can regrow parts which may be accidentally cut off, but the more complex the animal is, the less able it is to replace missing bits. If your cut is too big or too deep, the body will fail to repair it properly, the new cells are of a different type from the missing ones and show a permanent scar.

In most adult animals, some cells can divide for repairs, but there is no continued growth of the animal as a whole. Trees and some other plants keep growing as long as they live but their growth is confined to specialised growing points at the tip of the stems, the tip of the roots and within their outer tissues causing growth in girth. Even within species there may be short and tall individuals because size depends not only on environmental factors but also on genes (see p.101).

Cell division in a growing creature, or in the repair of damage, usually stops at the right point. Sometimes, though, the control fails and cells go wild and totally out of control. They multiply so that a large lump of tissue grows where it is not supposed to grow. This lump is called a tumour. It may be merely inconvenient, or it may be dangerously active, in which case it is called cancer.

Scientists have found that repeated exposure to X-rays or radioactive substances can stimulate cells to divide in this uncontrolled way. Some chemical compounds called carcinogens will cause cancer if they come in contact with living cells over long periods. For example, there are several kinds of carcinogens in products made from tobacco which cause lung cancer.

And then there were two

In 1859, Rudolph Virchow, a German physician published a research paper in which he wrote:

> *Where a cell exists there must have been a pre-existing cell, just as the animal arises only from an animal and a plant only from a plant. The principle is thus established, even though the strict proof has not yet been produced for every detail, that through the whole series of living forms, whether entire animal or plant organisms or their component parts, there are rules of eternal law and continuous development, that is, of continuous reproduction.*

He then coined one of the most meaningful and succinct phrases of cytology: '*omnis cellula e cellula*' – all cells from cells.

We take it for granted today that all cells result from cell division. The whole process can be demonstrated to classes of students in school laboratories with easily obtained materials and with the type of microscope usually available to schools. In the nineteenth century, however, things were quite different. It took some brilliant studies by the self-taught amateur, Wilhelm Friedrich Hofmeister to show how cells divide. He worked in his father's business in Leipzig as a publisher and bookseller. In his spare time, he was a pioneering microscopist. There are stories of his obsession with this interest, relating how he would rise at four in the morning to devote time to his studies using his microscope, before going to work. In 1848 he published his description of cells in the process of dividing, and although he attached no importance to his observation of certain rod-shaped bodies in the nucleus of stained cells, we may regard this as the discovery of what are known today as chromosomes. In fact, it was not until 1888 that the German anatomist, Wilhelm von Waldeyer, gave them their universally accepted name (from the ancient Greek *chromos* = colour and *soma* = body).

By 1880 there was general agreement that cells in both plants and animals are formed by equal division, with the nucleus always dividing before the rest of the cell. Some scientists thought that each chromosome divided in half across the middle. However, another gifted German biologist contributed to cytology in 1882. He was Walther Flemming and his work was outstanding for a variety of reasons. First he had access to the most sophisticated microscopes of his day because German technology in the field of precision optical engineering was second to none. Second, new dyes were being developed for staining biological material for study with the microscope and the combination of these two advances enabled Flemming to carry out work which would have been impossible only a generation before. He and his co-workers, Edouard Strasburger, and A. Q. Van Beneden, observed the lengthwise splitting of chromosomes during cell division and saw that the split halves are accurately distributed to the two new daughter cells. Thus the process of cell division in body cells was seen as an orderly series of changes. They called the process mitosis (from the ancient Greek *mitos* = thread).

Since the mid-twentieth century, dividing cells have been recognised as very busy places, swept by waves of biochemical activity. In our brief look at the shapeless *Amoeba* (see p.14), we saw that cells without nuclei do not divide. The centre of activity is the nucleus; so let us examine the nuclei of some dividing cells. The various stages are easiest seen in the tip of the root of garlic which can be obtained and grown at any time of the year.

To see what is going on in the nucleus, we must stain the cells with a special dye. In fact none is better than the one discovered by the German biochemist Robert Feulgen in 1924. He found that after warming the separated cells with strong acid he could colour chromosomes a brilliant crimson with a dye called fuchsin. With a normal school microscope the chromosomes show as a tangle of worm-like threads. With more sophisticated techniques involving micro-photography, and some sticking and pasting, it is possible to count the chromosomes in a cell. Garlic will have 16. In 1956, it was discovered that we have 46. The number of chromosomes is constant according to the species. A misconception that some students sometimes have is that the more complex the organism, the more chromosomes it has. This is not true; for example: *Amoeba*, one of the simplest of all animals has 50; a goldfish has 94; a dog has 78 and so does a chicken. Note that they are all even numbers. The significance of this fact will be seen when we look at genetics (see p.95).

As we look at chromosomes, we see that they exist in pairs in normal body cells. There are eight pairs in the cells of a garlic root tip. The chromosomes in your body can be sorted out and arranged in 22 matching pairs, with one pair of different ones. The odd pair are the sex chromosomes, of which there are two types called X and Y. Female mammals have two X chromosomes and males have an X and a Y. Strangely, the situation is reversed in birds, butterflies, and some fish.

Since chromosomes are made mainly from DNA (see p.101) and protein, the obvious way to see that each new cell formed by cell division has exactly the same amount of these chemicals, is to duplicate the chromosomes and distribute a full set of pairs to each new cell. This results in the duplication of the nucleus and is accompanied by a cleavage of the cytoplasm. These two processes normally take place concurrently and the result is a pair of cells, which are exact replicas of the original. Duplication of the nucleus and cleavage of the cytoplasm account for growth by cell division and also for the replacement of cells that have worn out or that have been killed by injury or disease. In cancer, cell division is uncontrolled and the chromosomes are grossly distorted. Some cells, particularly those of bacteria and some other simple plants (blue-green algae), do not have distinct nuclei. When their cells divide, the DNA becomes duplicated, but not in mitosis. Throughout the vast array of living organisms, cell division accompanied by mitosis is the form of cell reproduction most commonly found.

The dance of the chromosomes

By 1882, Flemming had distinguished nine phases in the process of mitosis. When the cell gets ready to divide, the membrane around the nucleus disintegrates. The chromosomes become visible as a tangle of stubbly spaghetti-like threads and closer examination shows that each one is really a double filament – in effect the number of chromosome pairs has doubled, although the twin chromosomes lie closely parallel to one another and are still attached to each other at one point.

But while we have been watching the nucleus, something else has been going on in the cytoplasm. Fine threads called spindle fibres have developed. At the point where they are joined, the fibres become attached to the twin chromosomes and in animal cells, definite pinching-in begins to develop at the edge of the cell as the cell membrane contracts.

Now comes the most dramatic phase of all. Each chromosome half begins to move away from its partner, or perhaps is pulled away by the contracting spindle fibres, to opposite sides of the cell. In the last stage, a new nuclear membrane forms around each group of chromosomes. There is now a new nucleus which is an exact duplicate of the original. Finally, the contracting cell membrane pinches through the middle of the cell and two blobs of life exist where before, there was only one.

The whole process, which resembles some well organised line dance, can occur in as little as half an hour and will normally take no more than a few hours. It is essentially the same in all animals and plants. However, in plants, the star-like structures that you see in animal-cell mitosis do not appear, and the cell does not pinch in two. Instead, a new cell wall grows across the old cell between the two new nuclei.

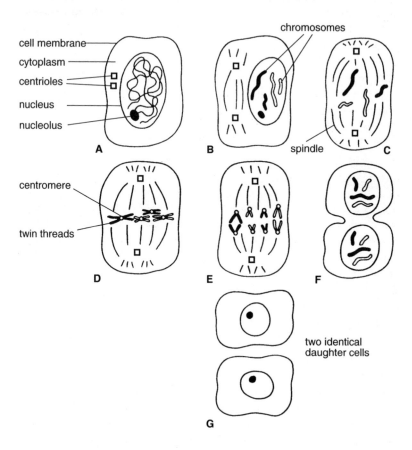

Figure 1.7 Cell division – mitosis

The bare necessities of life

The human body is about 65% water, 15% protein, 15% fatty materials called lipids, 5% inorganic materials, and less than 1%

carbohydrates. Proteins, the most complex of life substances, are found in, for example, muscle, hair, horn, silk and blood. Perhaps the most familiar protein to most of us is the white of an egg. Lipids in the body include substances similar to butter or lard. Everyday instances of carbohydrates are potato starch and the sugar you might use as a sweetener. During the early nineteenth century scientists were beginning to recognise that these three classes of organic substances – protein, lipids and carbohydrates – occur in all living things. In 1838, Gerardus Johannes Mulder decided that the protein was the most important of these and coined its name, which means 'primary' or 'holding first place'. In fact, it was the inveterate and famous word-coiner, Jons Jacob Berzalius who suggested the name to Mulder. Today, we know that he was correct. Proteins are, indeed, the class of chemicals that forms the basic role in life processes.

Is there a vital force?

In the early exploration of the chemistry of living things, biochemists, were handicapped by their belief that substances of organic origin had some mysterious power that made them different to inorganic substances. They called this 'power' a 'vital force'. Thus, one of the most well known scientists of all time, the Swedish chemist, Berzalius, is recorded as having said that life substances could never be made in the laboratory.

This belief in a vital force, known as 'vitalism', was widespread among chemists in the early nineteenth century. Perhaps it was the sheer complexity of the chemistry of life substances that convinced the vitalists that they were right.

In this context, an obscure German research student, Friedrich Wohler, made a discovery which would change current thinking completely. While working in Berzalius' laboratory in 1824, he was investigating cyanic acid, which contains nitrogen. As part of this research, he attempted to make the ammonium salt of the acid, ammonium cyanate. To his amazement, he made crystals of urea, which he knew was produced and excreted by animals. Surely, it was impossible for him to have made it in a test tube?

Perhaps, if his mentor, Berzalius, had not been one of the leading chemists of the day, and a firm advocate of vitalism, the young

Wohler would have discussed his puzzling discovery with him. As it was, he preferred to wait until he could double-check his findings. Leaving Berzalius' laboratory in Sweden, Wohler returned to his German homeland and became a teacher in Berlin. There, he confirmed his original observations and found that urea and ammonium cyanate contained the same chemicals joined together in the same proportions. Other famous chemists, the German, Justus von Liebig and the French Joseph Gay-Lussac, added the results of their research to Wohler's original discovery and came to the same conclusion. Legend has it that, in celebration of the confirmed marriage between the organic and inorganic worlds of chemistry, Wohler and Gay-Lussac waltzed around the laboratory together! Most unusual behaviour indeed for otherwise sedate members of nineteenth-century academia.

In 1828, Wohler was finally sure that he could make organic molecules from inorganic molecules and wrote to Berzalius, 'I must tell you that I can make urea without the use of kidneys, or of any animal, either man or dog'. The path was laid for the downfall of vitalism, although some chemists clung on to its appeal for many years.

Liebig and Wohler joined forces to explore the chemistry of other organic substances and were so confident of their way forward that, in 1838, they predicted the possible laboratory synthesis of all organic chemicals. Liebig also discredited the theory of vitalism, affirming that living things operated according to chemical laws. He claimed that there was no need to invoke any mysterious vital force to explain organic chemistry. As professor of Chemistry at Giessen University, he founded the first teaching laboratory. Many of Europe's most brilliant students flocked there as its reputation grew. His work influenced many of the greatest chemists of the nineteenth century. To understand life, insisted Liebig, it was first necessary to understand the power and subtlety of chemical action. The molecular biologists and genetic engineers of today would agree.

Giant molecules

Although chemists agreed on the prime importance of proteins, these substances are so complex that, at first, they could not make any headway in elucidating their structure. The earliest successes were not with proteins but with lipids. These contain carbon, hydrogen and oxygen only.

Michel Eugene Chevreul, a Frenchman who lived to the ripe old age of 103, began work on their structure as early as 1813. He showed that they contain a mixture of substances called triglycerides. These are compounds of glycerol (glycerine) and various acids called fatty acids. Glycerol is a member of the class of compounds called alcohols. By the 1850s another French scientist, Pierre Eugene Marcelin Bertholet has synthesised lipids in his laboratory, thereby adding another nail to the coffin of vitalism.

Like lipids, the carbohydrates contain only carbon, hydrogen and oxygen. However, their molecular structure remained a mystery until the 1870s, when the famous German scientist, Emil Fischer isolated various sugars and made the first steps towards working out their molecular structure. He succeeded in synthesising glucose and fructose as early as 1887. By 1883, Bernard Tollens suggested that the glucose molecule is made up as a ring structure and this was confirmed by the British scientist, Walter Norman Haworth in the 1920s and 30s. The simplest sugar is glucose. The sugar that some people add to their coffee is sucrose (cane sugar) which has a molecule, made of two units, glucose and fructose joined together. Sucrose is therefore called a disaccharide made up of two monosaccharide units, glucose and fructose, joined together. Chains of many sugars joined together are called polysaccharides, for example starch and cotton (cellulose).

Protein structure

The structure of proteins was first studied in detail by one of the most outstanding chemists of all time, the German, Emil Fischer. As long ago as 1819, a French chemist, M.H. Braconnet had isolated amino acids, the building bricks of proteins. Fischer showed how these units were linked together in chains joined by bonds called peptide linkages. Chains of amino acids joined in this way are called polypeptides (many peptides). It is known today that there are twenty amino acids making up all the known proteins by being arranged in various combinations. The number of ways that twenty units can be combined with repeated sequences is astronomical. A medium sized protein is made up of about 5000 atoms in total. These giant molecules were so difficult to analyse that progress was not really made until the 1940s when Charles Chibnall and his colleagues at Cambridge University attempted to work out their structure. Frederick Sanger spent ten years working out the amino acid sequence of insulin, a relatively small protein with 51 amino acids. He was awarded a Nobel prize for this work.

But why the difference?

While biologists have insisted that, in essence, all living things are the same, one cannot fail to be struck by the variety of types. Fish, chickens and apples are so different that our claim for similarity requires more explanation.

All forms of life, apart from viruses, are similar in that they are made of cells which are built on the same general pattern and which function in a similar way. We also find lipids, dissolved minerals, carbohydrates, and particular proteins in the cytoplasm which makes up the bulk of cells. However, anyone who has eaten fish, chicken or apples knows that there has to be something different about them. We also know that they grow in different ways and do quite different things. Compared to fish and chickens, apples do not appear to do much at all!

Most of the difference, of course, is chemical. First, the proportion of proteins, lipids and carbohydrates differs and so does the amount of water. The proteins differ too. They are made of different amino acid sequences. Each species and each individual within a species has its own special proteins.

We can eat fish or chicken because our digestive system breaks them down so that our cells can build up the protein that we need. As an analogy, let us pretend that an office building is demolished into its individual bricks. The office is a chicken protein molecule; the demolition is digestion; the bricks are the amino acids. Originally the bricks were arranged in a particular way so that the building looked like a familiar office block. The individual bricks could be taken somewhere else to build a completely different building with a completely different function – a fire station perhaps. In a similar way, amino acids are carried from the digestive system, in the blood, to cells where they are rearranged to make human protein. Or, if a fox had eaten the chicken, the fox's cells would rearrange the amino acids to make fox protein. So that is one reason why humans and foxes don't look like chickens, even though they might eat their protein! Incidentally, if whole chicken protein molecules were injected into our blood before being digested to their amino acid constituents, they would be rejected by the body's immune system. Indeed, large amounts of 'foreign' protein could kill us.

The second difference is in the chemicals needed for controlling cell activities. If cell chemistry is to work, the cells must be supplied with tiny amounts of chemicals known as vitamins. Plants make their own vitamins, but animals cannot make all that are necessary. Without vitamins, the cells in our bodies are not able to carry out their work, no matter how excellent the food supply may be otherwise.

Chemists in laboratories can break protein molecules apart by boiling them in weak acid solutions. Equally drastic treatment is needed to join amino acids together in test tubes. The wonder of living things is that they can do both at relatively low temperatures with the use of special proteins called enzymes.

There is one specific enzyme for every chemical reaction taking place in the body. Without vitamins, the enzymes will not work. Living things differ in their enzymes and in the vitamins needed to make them function properly. Fish and chickens, for example can make Vitamin C. Apples need it and make it. Humans need Vitamin C and cannot make it – we get some from apples.

If we lack certain vitamins, our bodies suffer from muscle diseases, skin disorders, nervous and bone diseases. Yet so small are the quantities needed that a balanced diet will supply all the body's needs. An overdose of certain vitamins can have dangerous side effects.

Vitamins receive so much publicity that we sometimes overlook another important chemical requirement; metals and other elements that occur in the soil and are commonly called minerals. All living things need sulphur and phosphorus. Animals with skeletons or shells need substantial amounts of calcium; iron is important in blood formation. Copper is needed for the blood of some animals. Sodium and potassium are needed for the correct balance of body fluids and nervous reactions. Plants need magnesium to make their green pigment, chlorophyll, without which all life on Earth would be impossible. Without chlorophyll, photosynthesis (see p.38) would not be possible – without plants, there would be no animals.

So while we can say that the bare necessities of life are the same, we are quite safe in saying at the same time that it is all wonderfully different.

How to cross a membrane

Everything the cell needs must pass through its membrane – the plasma membrane. Useful molecules must be retained in the cell, while waste products must pass out or else they would interfere with the chemical activities of the cell and become poisons. It is the cell membrane that keeps the cell's integrity and retains cytoplasm. In fact, if we did not have cell membranes, we would be just a runny mass of jelly on the floor! A membrane must not hinder the passage of all molecules – only some. So cell membranes are selectively permeable or differentially permeable.

Anyone who looks at an animal cell, even with the highest magnification available with an ordinary light microscope, will see the cell membrane as a boring looking thin line surrounding the rest of the cell. It is so thin that its presence was first postulated on circumstantial evidence alone, until the advent of the electron microscope and techniques for studying molecular shapes and sizes. Research in cytology (the study of cells) has shown that plasma membranes are made of two layers of lipid-phosphate molecules (called phospholipids) in which are scattered various proteins (see p.9).

The image is called the fluid mosaic model because the membrane has the fluid consistency of butter on a warm day, rather than the solid nature of lard. The term 'mosaic' refers to the fact that the proteins are scattered about as in a pavement mosaic and can move about. An analogy would be floating icebergs in a sea of oil. Some membrane proteins can recognise specific materials and either allow them to pass through or keep them out. The unfortunate consequences of a malfunction of this are seen in people who suffer from cystic fibrosis which is the most common inherited disorder in humans. The problem is the inability of cell membranes to regulate the correct amount of salt passing in and out of cells.

Cells have membranes with their own specific proteins. For example, your red blood cells have special protein 'labels' on their membranes which determine your blood group as A, B, AB or O. Other molecular labels determine acceptance or rejection of a transplanted kidney, heart, or liver.

The functions of cell membranes are aided by some fundamental physical characteristics of matter. Molecules are always in motion. The molecules of solids vibrate back and forth. When solids (solutes) are dissolved in liquids, (solvents), the motion of the solvent molecules knocks them around randomly. Due to their constant motion, solvent molecules and gas molecules bump into each other frequently and, as a result, will spread out. That is why a little perfume goes a long way. Scientists call this process diffusion.

As an example of how the process works, let us consider the single-celled animals like *Amoeba* again. Here, molecules of oxygen dissolved in water diffuse through tiny holes in the membrane. As the cells use up the oxygen and produce carbon dioxide in the process of respiration, the concentration of carbon dioxide inside becomes greater than that on the outside and so there is a net flow outwards down a concentration gradient. Now the oxygen is again lower in concentration on the inside than it is on the outside and so there is a net flow of oxygen inwards down a concentration gradient.

In larger animals, water or air is made to flow over the gills or into the air sacs of lungs, which have a large area of tissue well supplied with blood. Oxygen diffuses across a single layer of cells into the blood which carries it all over the body and brings back carbon dioxide to diffuse outwards across the breathing surface.

Other small molecules behave in the same way when they are dissolved in water and diffuse through small holes in membranes. Glucose and other simple sugars get through, and so do amino acids. Waste products like urea can leave cells in the same way. The absorption of glucose molecules across the membranes of some cells is facilitated by the use of some protein molecules which act as carriers.

Large molecules are simply stopped; but chemicals like ether and other fat solvents, enter cells easily, even though their molecules are too large to pass through the holes. Such molecules accumulate in the lipid component of the membrane and interfere with the cell's normal routine. That is why ether and other anaesthetics alter the way in which our nervous systems normally carry nerve impulses. It also explains why solvent abuse is often fatal in its consequences.

Unless conditions on both sides of the membrane are exactly balanced, water will rush either in or out, down a concentration gradient, causing cells to swell up and burst, or shrivel up and die. Diffusion of water across a selectively permeable membrane is called osmosis.

So far, all of the methods of crossing membranes described have relied on passive movement of molecules down a diffusion gradient. However, there are times when molecules pass into cells against a concentration gradient. In this case, the cell must expend energy and use a carrier molecule to 'pump' materials up the gradient. It is called active transport. For example, sodium is pumped out of cells up a concentration gradient and potassium is pumped into cells against a concentration gradient.

Glow-worms and cramp fish

There is a constant activity in living things which requires a constant supply of energy. Hearts beat, muscles twitch and stomachs digest food. Flowering plants spend all their lives rooted in one spot but are also very active. They grow taller and stouter, blossoms open and close, stems twist and turn towards the sun, roots reach out towards water. Even where there is no obvious motion, microscopes reveal that the cytoplasm of the cells is always churning around. Besides motion, cell activities produce heat when they release energy. Small electrical changes occur in our nerve cells and brains. Electric fish produce great charges of electricity. In fact, the first treatment by electrotherapy took place with the use of a type of electric fish from the Mediterranean.

Wealthy Romans of classical times were carried to rock pools on the shore where they would be encouraged to put their feet on the electric ray, *Torpedo*, for treatment for their gout. In the seventeenth century, the English name for *Torpedo* was cramp fish. In the 1960s, thousands of years after ancient races discovered this 'living electricity', visitors to the Naples aquarium were invited to touch a small *Torpedo* in a tank, to receive a mild electric shock. After being touched by a few hundred visitors each day, the poor old fish looked a little worn out and weary!

Glow-worms and weird and fascinating animals of the oceans' abyss release energy as light. Motion, heat, and electricity are all forms of energy. The cells of living things use energy and convert it from one form to another; but they cannot create it. Life's energy comes from fuel – the food that plants can make by photosynthesis. Energy is the 'push' in nature that makes things happen; or as physicists say, 'energy is the ability to do work'. It can be classified in two ways. If you wind up the spring motor of a clock or clockwork toy, the tightly wound spring has *stored*, or *potential*, energy in it. As the spring begins to unwind and turn wheels and gears, the moving, active energy is called kinetic energy.

When carbon atoms are joined in chains or in rings, their bonds are a form of potential energy. The bonds are similar to wound-up springs with their stored energy.

A log burning slowly in a fireplace is undergoing a chemical change in which energy, stored as chemical bonds, is being released. The cellulose in the cells making up the log breaks down to carbon dioxide and water, and the bond energy is rapidly converted to heat and light. Animals also give off carbon dioxide and water in a process similar to burning. Although the end products of the two processes are the same, cells 'burn' their fuel in quite a different way. Instead of letting energy escape rapidly, cells must save it to be used for many purposes. So chemical reactions in cells take place in small steps that release energy slowly.

There is a transfer of energy in every chemical reaction. Some reactions release energy and others store it up. In cells these two types of reactions are linked by special chemical compounds that store energy and then release it when chemical bonds are broken. This concept was first suggested in 1941 by the German-born American chemist, Fritz Albert Lipmann. The most important of these compounds contains three groups of phosphate (a combination of phosphorus and oxygen). It is called adenosine triphosphate or ATP for short. ATP is a kind of chemical storage battery. In the ATP molecule, the three phosphate groups form a chain, hooked to the rest of the molecule.

A --P--P--P

where A = adenosine and P = phosphate.

The most important point about ATP is that the last phosphate group and bond can easily be separated from the rest of the ATP molecule and joined to other compounds. The loss of one phosphate group (ion) converts ATP into adenosine diphosphate (di meaning two).

$$A\text{--}P\text{--}P\text{--}P \text{ ----------> } A\text{--}P\text{--}P + P$$
$$\quad\text{ATP} \qquad\qquad\qquad \text{ADP} \qquad \text{phosphate}$$

The molecule that receives the phosphate group from ATP can either use the energy for some chemical reaction, such as linking amino acids together, making muscles contract or lighting a glow-worm's tail; or it can pass it on again. Sooner or later however, the energy in the phosphate bond will be used and the phosphate will be released. When this happens in the presence of some other energy-releasing reaction, the phosphate group will join an ADP molecule and convert it back to ATP, ready to start the cycle all over again. Incidentally, ADP is always present in living cells. Indeed, when space probes are sent to test for life on another planet, robot sensors carry out tests on the surface of the planet to detect the presence of ADP to confirm the presence of life as we know it.

ATP is the energy storage battery of all cells, but it can drive reactions in two directions. In plant cells which make their own food from carbon dioxide, water and soil minerals, ATP supplies the energy to join the carbon atoms into chains of carbohydrates. Animal cells cannot make either carbohydrate or all the amino acids they need. They depend on plants. The ATP in animal cells is used to split carbohydrate molecules to obtain their potential energy for use by the cells. For his key discovery, Lipmann shared the Nobel Prize in physiology and medicine in 1953.

The cool fire of life

Glucose is the fuel for cells. Its energy is made available by respiration; the slow, stepwise 'burning' that one of the founders of modern chemistry, Antoine Laurent Lavoisier (1743–1794), thought was similar to burning wood. The chemical changes that take place in respiration are often described in an over-simplified chemical equation:

$$C_6H_{12}O_6 + 6O_2 \text{ ----> } 6CO_2 + 6H_2O + E$$

or, in words: glucose plus oxygen equals carbon dioxide plus water plus energy.

This chemical sentence, however, is simply a statement of reactants and products. It certainly does not explain the complex chemistry involved in a series of stages and does not give any idea of the form in which energy is released.

Although it can be made to work by burning sugar on a fire, it tells very little about what actually happens when it takes place in every single living cell which exists on Earth. Surprisingly, the first few steps in the process do not require oxygen, but occur anaerobically (without oxygen). It is the basis of fermentation which is the method by which yeast converts sugar to alcohol. In 1905, two British chemists, Arthur Harden and William John Young, discovered that the reaction begins when the six-carbon molecule, glucose, reacts with two phosphate molecules. This first step results in the transfer of two phosphate groups to glucose. By the 1940s, it was realised that the source of the phosphate groups was ATP (see p.32). The six-carbon, glucose-phosphate combination splits into two, three-carbon molecules. Each of these join with another phosphate group; not one from ATP, but from other phosphate compounds dissolved in the cell's fluids. So far in this explanation of energy release, the cell has gained no energy and has lost two ATP molecules. In the very next reaction, however, the cell starts to make a profit.

Each of the three-carbon compounds releases energy by giving a phosphate to an ADP molecule to make an ATP molecule. After a few more minor changes, another such reaction transfers more energy and the two remaining phosphate groups to two more ADP molecules, converting them to ATPs.

Although it required two ATP molecules for each glucose molecule to begin this series of reactions, four ATP molecules result. You must admit that a hundred percent profit is a good return on the original investment of energy!

But what about the oxygen in the original simple equation? Where does it come in to the reaction? The fact is that respiration can now go in one of two ways. It can continue without oxygen and end in the production of ethanol (alcohol) and carbon dioxide. This is what happens in simple organisms like yeast, when the change from

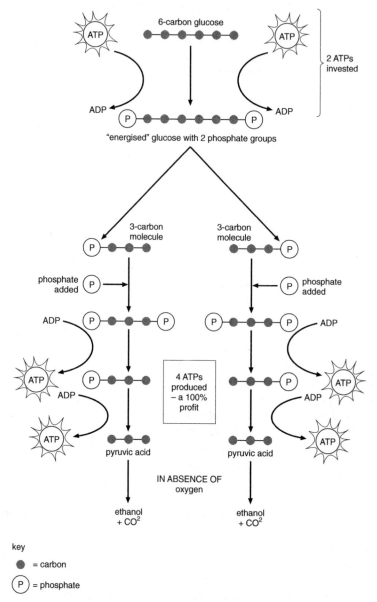

key

● = carbon

(P) = phosphate

Figure 1.8 Summary of anaerobic respiration

sugar to ethanol is called fermentation. This, of course, is the basis of one of the earliest forms of biotechnology – the production of all known alcoholic drinks and of bread.

However, a yield of only four ATP molecules per glucose molecule does not provide enough energy to satisfy the needs of more complex organisms.

Furthermore, it would be wasteful not to use the energy still locked in the two, three-carbon molecules. From this point, oxygen makes its entry to the scene and the word 'oxidation' is used to describe the chemical changes that follow. The cells of our bodies and other air-breathing animals complete respiration in a series of reactions involving oxygen. These complicated reactions begin by pruning off one more carbon atom to make a two-carbon unit. Now follows a cycle of reactions that first lengthen the chain by building the two-carbon units into six-carbon molecules, and then shorten the newly-formed six-carbon chains to four again. At each step, hydrogen atoms are removed and eventually join with oxygen to form water. The necessary removal of hydrogen to build six-carbon chains releases energy which is stored in ATP molecules. In other words, removal of hydrogen (oxidation) re-distributes the energy within the glucose molecule and makes it available to be stored in ATP.

Still more energy is released in the cycle that again shortens the length of the carbon chain from six atoms to four. The two carbon atoms that are lopped off in this part of the cycle are the source of carbon dioxide that we breathe out, shown in the original equation. In all, 38 ATP molecules result from the energy contained in the single molecule of glucose with which we started. More than 55% of the potential energy in the glucose bonds has been converted to useful kinetic energy, an efficiency that is higher than that of most car engines.

A series of chemical reactions so closely coupled must surely require that the compounds stay close together. In 1948 it was discovered that cell respiration involving oxygen takes place in the mitochondria (see p.10), the tiny sausage-shaped compartments in the cytoplasm. Some of the energy transferred to ATP is used by the mitochondria to start respiration, but most of the ATP passes out into the cell for other activities. The mitochondria supply the cell

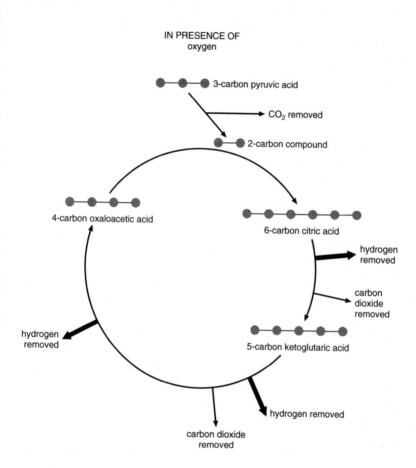

Figure 1.9   Summary of aerobic respiration

with ATP and are rightly called the 'powerhouses of life'. Although mitochondria were first detected in 1898 by the German biologist, C. Benda, he did not link them with any physiological process and, in fact, thought they were 'threads of cartilage' – hence the derivation of their name from the ancient Greek roots. It is now known that the ATP released in them can be used for all energy-using purposes and is aptly called the currency of the cell.

Light and life

Every day, year in and year out, century upon century, living things 'burn' up millions of tonnes of sugar. Where does it all come from? Chemists in a well equipped laboratory can produce a few crystals of sugar after many hours and much expense; but while the chemists work at the unrewarding task, hooking one carbon atom to another in a series of complex reactions, the trees outside their laboratories' windows produce kilos of sugar and use it to make other carbohydrates, fats, vitamins and protein.

In the cells of green leaves, carbon dioxide and water join to form sugar in a series of reactions that appear to be just the reverse of respiration:

$$6CO_2 + 12H_2O \text{ --------> } C_6H_{12}O_6 + 6O_2 + 6H_2O$$

The energy required comes from sunlight and the name for the process is photosynthesis. The leaves' remarkable energy-capturing power holds the key to life. On the 'cost efficiency' side, a leaf can change about 35% of light energy that falls on it to chemical energy. The leaf is an organ, and within it we see a variety of cells organised into tissues. The top and bottom are protected by irregular, flat epidermal cells that make up the 'skin', or epidermis. The lower epidermis has many tiny openings, the stomata, through which gases and water pass in and out. Each stoma is made of a pair of guard cells that allow the pore to open and close like a mouth. Indeed, the name, *stoma* comes from the Greek for mouth.

Between the two layers of epidermis the leaf is packed with layers of cells. Uppermost is a layer of rather long cells quite closely packed, known as palisade cells. Just below is a layer of loosely-packed, fat, roundish cells called the spongy layer. Gases – carbon dioxide and oxygen – are stored in this layer. Both types of cell are fairly typical in structure except for their cytoplasm which contains many packets of the green pigment, chlorophyll.

It is rather surprising to find, perhaps, that chloropyll is contained in these chloroplasts rather than being diffused throughout the cell.

Running down the centre of the leaf are veins made up of conducting tubes and woody material. The veins have two important jobs. Their rigidity serves as a support to keep the leaf spread out so that a maximum of their surface is exposed to light. The second job is similar to that of veins in our own bodies; the veins bring water to the cells and carry away manufactured foods to all parts of the plant.

The idea that green leaves are food factories developed slowly. The complete story was not known until the middle of the twentieth century. It all began in 1610 when Johann Baptista van Helmont (1577–1644), a Flemish doctor and a pioneer of physiology, planted a willow branch in a barrel of soil. In five years the branch grew into a small tree that weighed 169 pounds. It had weighed five pounds when van Helmont planted it and had used up only a few ounces of soil. He concluded that most of the tree's weight came from water.

In his words, … 'therefore one hundred and sixty-four pounds of Wood, Bark, and Roots arose out of water only …'.

In 1772 Joseph Priestly (1733–1804), the English chemist who discovered oxygen, found that plants could, as he put it, 'restore air' that had been 'spoiled' by burning candles. Today we know that plants give off oxygen during photosynthesis that replaces the oxygen used up by burning candles. Priestly knew only that when one candle had been burned in a jar until it went out, no more candles would burn in the same air. As far as he was concerned the air was 'spoiled', but after a plant had been introduced, the candle could be relighted. He published his work in *Experiments and Observations on Different Kinds of Air* in 1774.

At this point ideas developed rapidly. In 1779, Jan Ingenhousz (1730–1799), a Dutch doctor and chemist, found that plants corrected 'spoiled' air only when sunlight shone on them. In fact, he demonstrated that the green parts of plants perform the process

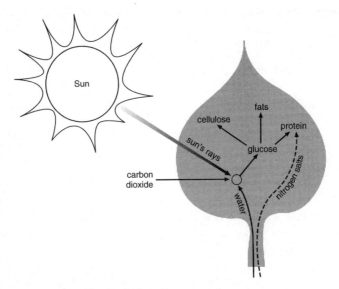

Figure 1.10   Effect of the Sun on plants

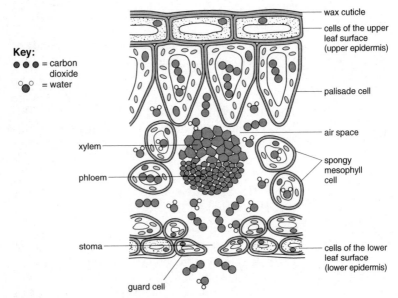

Key:
● ● ● = carbon dioxide
○○ = water

Figure 1.11   Structure of a leaf

we now know as photosynthesis. Indeed, he had observed chlorophyll, though he did not name or define it. Two French chemists, Pierre Joseph Pelletier and Joseph Bienaime Caventou, in 1817 were to isolate this, arguably, the most important pigment in the living world. They named it chlorophyll, from the Greek words meaning 'green leaf'. Then in 1865, the German botanist, Julius von Sachs showed that chlorophyll is confined to little containers which he called chloroplasts.

Ingenhousz showed that plants give off oxygen in the presence of sunlight. In 1782, Jean Senebeir, a Swiss minister with a passion for scientific experiments, found that oxygen production in green leaves stops, even in sunlight, unless the air contains carbon dioxide. By 1845, Julius Mayer, a German chemist, was able to say that the 'plant world forms a reservoir in which – the sun's rays are fixed and … lade down for future use'.

When the electron microscope became available in the 1950s and subsequent discoveries of the cell's details were combined with the biochemist's discoveries, it was found that photosynthesis is a two-step process that occurs inside chloroplasts. The first part, truly photosynthetic, takes place in light. Here the energy is captured by the chlorophyll. In the second step, the captured energy is used to join carbon atoms in the glucose molecule.

How to catch a sunbeam

When the sun makes the horizon blush on a bright spring morning, its rays fall on the mantle of green vegetation that covers much of the land on our planet and also on a floating blanket of plant plankton in the sea. The energy in the light is in packages called photons, like infinitesimal bullets of light. When sunlight is passed through a glass prism it separates into a rainbow of colours as the photons are sorted out according to the energy they have. Chlorophyll, the green pigment in the leaves of plants, absorbs the photons of red and blue light best. When a photon of red light strikes the chlorophyll molecule, it is absorbed into the molecule. The photon has been 'captured' and the energy-packed chlorophyll has been 'excited', or 'activated'. This causes a small electrical charge to go around inside the chloroplast.

Now the trapped light energy from the sun is used to make ATP from ADP (see p.32) and phosphate groups in the cell sap.

As so often happens in nature, a second route has been provided to achieve the same end. In many plants there is another energy-storing compound called nicotinamide adenine dinucleotide phosphate. This is quite a tongue-twister, even for biochemists who often talk in these terms, so they abbreviate it to NADP. The bulk of the NADP molecule resembles ATP, but its active part is different. ATP has at its end a phosphate bond, consisting of phosphorus and oxygen linked together as if by a 'coiled spring' of potential chemical energy. The active end of NADP is a nitrogen-hydrogen combination in amino acids. It can pick up the electrical energy flowing inside the chloroplasts along with another atom of hydrogen and become $NADPH_2$.

The hydrogen, of course, comes from water, H_2O. When water is stripped of its hydrogen atoms, nothing is left over but oxygen, which returns to the atmosphere through the stomata of the leaf. We might guess that some of the hydrogen in $NADPH_2$ is handed on to carbon dioxide molecules, in order to convert them to CH_2O molecules.

At this point the light phase of photosynthesis is complete. The remaining dark phase begins. During the dark phase, ATP and $NADPH_2$ are used up. ATP provides the energy to join carbon dioxide into three-carbon chains, and $NADPH_2$ supplies the hydrogen atoms and energy to complete the molecule. Eventually the three-carbon compounds are linked to form the six-carbon chains of glucose molecules.

At least a portion of the glucose must be used up on the spot to provide energy for the cell, to keep its protoplasm in good repair, to manufacture protein and to push substances across the membrane of the cell. Excess glucose is made into starch (a long chain of glucose molecules hooked end-to-end to form rings) and kept in the cells for future use. In autumn, glucose reacts with leaf pigments to produce the rich reds, russets and gold that crown hardwood trees when their leaves begin to die. Possibly the only time when there is beauty in death.

Most of the sugar produced is carried away to feed roots and stems and to provide the energy and substance for growth, because it can

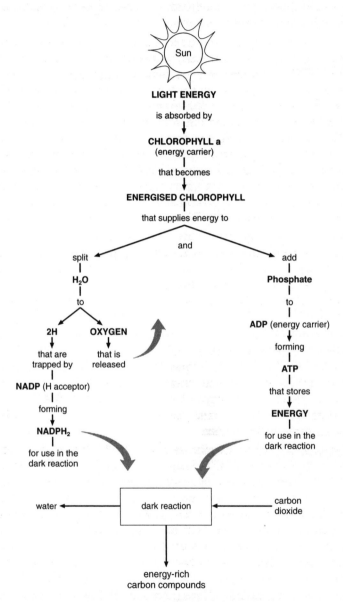

Figure 1.12   Summary of photosynthesis

be converted into fats or modified by the incorporation of nitrogen (from nitrates in the soil) to form proteins.

Last, but not least, the by-product of photosynthesis, oxygen, is every bit as important to all higher organisms as glucose itself. We could not live without oxygen; and without plants to restore 'spoiled' air, our oxygen supply would gradually be converted to carbon dioxide as a result of respiration. Most scientists who have studied the problem believe that when the Earth was formed and finally became cool enough to support living organisms, there was no free oxygen in the atmosphere. They guess that life on Earth began some four thousand, million years ago and that the first forms of life had to make do without oxygen. Gradually photosynthetic plants evolved. In the course of many hundreds of millions of years, these plants released the oxygen found in our atmosphere today.

So now we have come a full circle. We started with the sun and we end with it. Without our bright star there would be no possibility of life on Earth.

From root to leaf and back again

Plants produce food for the world of living things from two common substances, carbon dioxide and water. The carbon dioxide enters through the leaves, but the water and essential minerals enter at the opposite end, through the roots. Sugar, which supplies the plant's energy as well as all its other chemical needs, must be carried away from the leaves and distributed.

In the seed-producing plants, water and dissolved minerals from the soil are absorbed by the roots and pass along thousands of cells before getting to the root's central core which has hollow, drain-pipe shaped cells to carry materials in solution up to the stem and leaves.

These long, hollow cells in roots, stems and leaves are dead and have lost their protoplasm. They have firm woody walls which are permeable to water. They are called xylem (pronounced zeye-lem) and are surrounded by a ring of still another type of long cells. Sap containing the sugar made in the leaves passes downward through these outer tubes, called phloem (pronounced flow-em), to all parts of the plant.

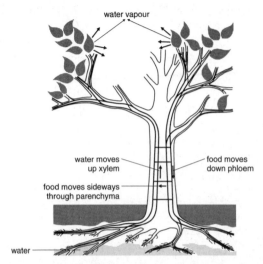

water vapour

water moves
up xylem

food moves
down phloem

food moves sideways
through parenchyma

water

Figure 1.13   How things move through plants

In the eighteenth century an English naturalist, Stephen Hales, carried out an interesting experiment. He removed a ring of bark from the outside of a growing shrub and cut away the layer containing the outer phloem; but he left the inner layer of wood. Water continued to pass up through the woody tubes so that the shrub did not wilt, but because the phloem had been taken away, food manufactured in the leaves, could not pass downward. After a few weeks, Hales noticed that a thickening had developed above the missing ring of bark, indicating that the food that was unable to pass downward had accumulated.

In the trunks of trees and the stems of other seed-producing plants, the two layers of conducting tubes, xylem and phloem, are separated by a layer of growing, dividing cells, called cambium. The cells of the cambium form layers of conducting cells on each side. The yearly production of woody tubes form the annual 'growth rings' that we see when the trunk of a tree is cut across.

This remarkable tissue is responsible for changing the diameter of a seedling of a giant redwood tree from a few millimetres, when it germinates, to the size that allows you to drive a car through when it is fully grown!

The stems of grasses, cereal crops and daffodil-like flowers contain bundles of both kinds of conducting tubes which run between the roots and leaves. There is no cambium in these, so once the stem has formed, it usually does not grow any thicker.

Solutions of manufactured foods are moved through the phloem cells of plants by pressure from the cells that take part in photosynthesis. These cells contain a lot of dissolved sugar, and so have a low water potential and therefore absorb water, which makes them turgid or swollen with water. This 'turgor' pressure forces the solution of food out through the phloem tubes to other cells in the same way that you can blow air through a straw.

The flat construction of a leaf exposes large surfaces to the air. This large surface area means that each leaf loses water by evaporation. This loss of water by plants is called transpiration. A large tree in full leaf has hundreds of square metres exposed, and may absorb and transpire up to half of its weight in water every day. An enormous amount of work is required to raise this amount of water to the top of a tree. A hectare of corn needs about five million litres of water from the time it germinates until the time it is harvested.

There is evidence to suggest that high summer temperatures and an increase in carbon dioxide in the atmosphere due to excess burning of fossil fuels, cause stomata to close. This reduces transpiration, which in turn, prevents the recycling of water. There has been a reduction of transpiration by about 3–4% since 1950 and this could be enough to influence the weather by making it drier.

When cells close to air, spaces in the leaf lose water to the air, osmosis (see p.31) causes water to diffuse in from neighbouring cells, and then from the water-conducting cells in the nearest leaf vein. In the xylem are continuous, tall, extremely thin columns of water stretching back to the roots. The osmotic 'suck' of the leaf cells pulls on the columns of water and lifts them. Thus the roots absorb more water to make up for that which is lost from the surface of the leaves.

In effect, the sun does the work of lifting the water by evaporating water from the leaves; and the water in root cells creates a positive pressure that pushes the life-giving liquid upward. At night, the stomata close; transpiration becomes very slow, and the light phase of photosynthesis ceases (see p.41). The plant tissues continue to use food molecules slowly, and therefore to use oxygen; and they give off carbon dioxide just as animal cells do even when the animal is asleep.

Food processing

An outstanding characteristic of living things is that they survive by taking in substances and either use them to make food, if they are plants, or they use them as food if they are animals. As Walter de la Mare observed:

It is a very odd thing –
As odd as can be –
That whatever Miss T. eats
Turns into Miss T.

You may be surprised to know that, biologically speaking, after you have finished a good dinner, your meal is still on the outside of your body. At least it is on one of your body surfaces, since, for all practical purposes, your digestive system is nothing but a tube with attached glands. Before the food can begin to satisfy your demands for protein, carbohydrate, fat, minerals, and vitamins, your meal has to get inside your cells.

In order to be absorbed for use by living cells, digestive enzymes must break down larger molecules to smaller, water-soluble molecules. Carbohydrates must all become glucose (the simplest sugar). Proteins must be digested to amino acids and fats to fatty acids and glycerol. These small molecules can then be absorbed. They pass across cell membranes into the cytoplasm of the cell where they are acted on by different collections of enzymes which either release energy from them or help build them into larger molecules.

When biochemists attempt to simulate the digestive process in test tubes using only inorganic materials, they find that they must boil the food in acids or other corrosive chemicals to achieve the same results as our digestive systems manage to achieve at body temperature with very innocuous chemicals. If, however, they add to their mixtures some digestive enzymes, test tube digestion proceeds quite nicely at room temperature or perhaps just a little warmer. Use is made of this in the production of biological washing powders, where digestive enzymes are used to digest stains produced by sauces, wine, fat, and such like. They are recommended for use at 40°C and, unless you happen to be allergic to them, will not affect your skin if they are handled. Arguably, enzymes are the most important chemicals in your body. A common misconception is that all enzymes break things down. This is not so. There are thousands of chemical reactions taking place in your body at this instant, all requiring enzymes. Most enzymes have nothing whatsoever to do with digestion. The majority are concerned with releasing energy in cells and in building up larger molecules from small molecules. There is just one relatively small group which helps the breakdown of food.

Perhaps it is not surprising that the misconception is so common because, the earliest experiments with enzymes involved those concerned with digestion. Also, the basic experiments on enzymes in school science lessons are often confined to digestive enzymes. In the latter part of the eighteenth century, Lazaro Spallanzani (1729–1799) carried out some very innovative investigations. He fed some hawks with meat enclosed in little wire capsules. After the birds had had time to digest the meat, he made them regurgitate the capsules and found that the meat had disappeared. Since it could not have been ground up by the birds, it had evidently been liquified by some agent in the stomach juices. In 1836, the German naturalist, Theodor Schwann named the stomach enzyme, pepsin, from the Greek word meaning to 'digest'.

Enzymes are biological catalysts. In other words, they are proteins made by living things which speed up chemical reactions and remain unaltered after the reactions. They do not provide any of the atoms that go into the reaction or which come out of it. When the reaction is complete, the enzyme can be recycled. Enzymes function very much like a lock and key. Each enzyme has a unique arrangement of molecules that gives it a particular shape. The chemical upon which the enzyme acts, (the substrate) acts like a key and fits into the particular shape of the enzyme molecule, which acts like a lock. When the two are combined to give an enzyme-

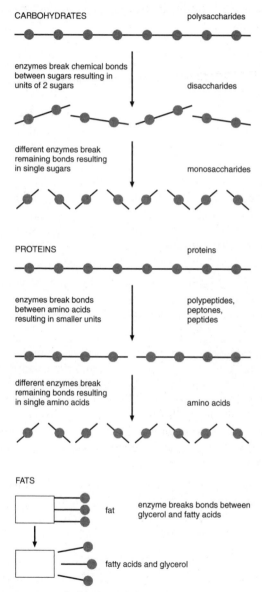

Figure 1.14 Breakdown of foods

substrate complex, the reaction takes place and the product plus the enzyme remains.

When applied to the digestive enzymes in our digestive system for example, a large molecule would combine with a particular enzyme and would then become a smaller molecule, leaving the enzyme to react again. The heat of our bodies, water, and some dilute acid or alkali from the glands, are also essential for the digestion to take place. The principle holds for all digestion taking place in all types of animals – from the simplest to the most complex.

Amoeba surrounds its food and forms a bubble (vacuole) separated from the cytoplasm by a membrane. Enzymes are secreted by the cytoplasm and diffuse into the vacuole where digestion occurs. The small products of digestion diffuse through the membrane into the cytoplasm. This is the most primitive form of digestion i.e. within a cell (intracellular) and is not sufficient for the processing of food in those animals that 'eat' relatively large pieces.

The cells of fungi, bacteria and those animals with digestive systems pour enzymes on to their food, digest it, and then absorb the products. This method is called extracellular digestion.

Jellyfish and many flat worms and roundworms carry out extracellular digestion in little more than a sac. Food enters the sac, digestive enzymes are poured on to it, and the undigested remains leave through the same opening. True digestive systems evolved first in the group of worms to which the earthworm belongs. This group is somewhat more advanced than the more primitive flatworms (tapeworms) and roundworms. An earthworm has a tubular digestive system with two openings – one at the head and one at the tail.

It is not all that surprising that the parts of the digestive system are often quite different in form in various animal groups because they have to be adapted to dealing with such a variety of foods. However, many of the parts are alike in function. For this reason they are given the same name. The chief changes between an earthworm and a human are that the digestive part of the human system is lengthened to house a series of glands that secrete different digestive enzymes, and the intestine is modified to provide more absorbing surface. The blood system, in order to transport

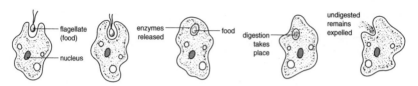

FOOD VACUOLE FORMING IN AMOEBA

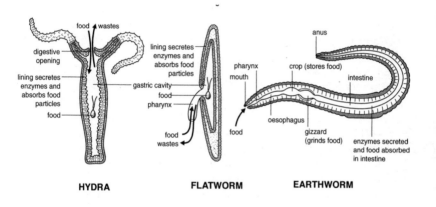

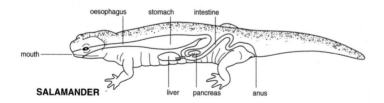

Figure 1.15   Digestive systems

soluble products of digestion becomes more elaborate. A liver has developed – a vast workshop where food molecules are stored, rearranged, built into new molecules, prepared for energy-releasing chemical reactions or excretion, and so on.

Food is prepared for digestion in many ways. The earthworm and chicken grind their food in a muscular gizzard containing small stones. Lobsters, crabs and some insects are messy eaters. They use mouth parts like knives and forks, shredding and cutting before swallowing. The starfish turns its stomach inside out onto its food and digests it before swallowing. Snakes swallow their prey whole and digest it over a long period at leisure. Fish and reptiles tend to have teeth adapted for trapping prey, whereas most mammals have teeth adapted for biting and grinding; exceptions are the filter-feeding whales.

Digestion in the cud-chewing animals (ruminants – cows, sheep, goats, camels, deer and antelopes) differs from that of most creatures because the food first goes into the rumen, one section of a four-chambered stomach. Afterwards, they return the food to the mouth and 'chew the cud'. In the rumen the food is acted on by some specialist bacteria which can digest the cellulose of plant cell walls. It is remarkable that, despite millions of years of evolution and the fact that ruminants only feed on plants, none have evolved a means of producing a cellulose-splitting enzyme. They all rely on bacteria and other microbes which live in their guts in a way that offers mutual benefit – the microbes have warm, moist, food-filled place to live and the mammals have cellulose digested for them, making this type of carbohydrate accessible to them as a source of energy.

Termites thrive in warm climates and feed chiefly on the supporting timbers of houses – a type of food even harder to digest than grass. They have specialist one-celled animals in their intestines to help them digest the cellulose in wood.

Similar bacteria live in the large intestine of the horse and other non-ruminant herbivores. We cannot digest cellulose, so part of a vegetarian meal is useless to us as a source of energy. In our case, cellulose acts as fibre and bulk for the intestinal muscles and helps to prevent constipation. The bacteria which live in our intestine do not harm us. Some make vitamins which we can use.

Exchange in simple animals

Animals that are made of many cells have many advantages over single-celled animals. The most obvious is adaptability. Humans, for example can live in a variety of climates and they can move if the weather in one place does not suit them. Animal migration is a common response of many-celled animals seeking situations that are more satisfactory. However, there are some disadvantages. A basic problem that goes hand-in-hand with being made of many cells, is how to maintain the life of cells that are far removed from the source of supply of life-giving materials.

A squirrel's tail, for instance, is more than an elegant decoration. It is a balancing organ, and without it, the squirrel's hair-raising leaps would be impossible. It is also a fact that a squirrel's tail is far away from its stomach where its food is digested. Digesting food is one problem; getting the products of digestion to the tail is another. Useful materials must be carried to cells where they are needed – as in the tip of the tail – and waste products must be carried away.

A simple cell does not need a transport system as long as it is small enough to receive food and water by diffusion, through its membrane. In one sense, the cell's entire surroundings function as the transport system.

The simplest of many-celled animals live in water. In jellyfish and their relatives, for example, water and food particles are taken into a simple digestive sac surrounded by a body made of only two layers of cells. Oxygen can reach both outside and inside the animal. Food is digested by cells lining the digestive sac and the products of digestion diffuse to other cells. Carbon dioxide produced in the inner and outer layers either diffuse directly into the water or indirectly through a series of water-filled tubes in the larger jellyfish.

If diffusion is to continue to work to maintain the necessary two-way traffic, the water in contact with the inner and outer surfaces of the creature must be renewed quite often. The movement of water may be helped by the beating of many hair-like cilia, or by the muscular movements of the tiny animal's body. Diffusion can also be helped by increasing the surface area in contact with the water. In relatives of the jellyfish, the digestive cavity extends out into the tentacles.

Imagine an animal with three layers of cells – the next step up in complexity from jellyfish. This poor creature's middle layer is so thick that food, digested in the central cavity, cannot diffuse to all of the cells. Oxygen cannot reach the central cells from the outside quickly enough to supply their needs; carbon dioxide cannot diffuse directly away from them. In fact, the cells in the middle layer would be starved, suffocated and poisoned, all at once.

There are two ways of avoiding this fate. Several hundred kinds of marine and freshwater worms known as flatworms continue to depend on diffusion. Because of their ribbon shape and the branching of their digestive cavities to provide more surface area for diffusion, they have very large surface areas compared with their bulk. The middle layer is so thin that it can all be supplied. The second way, appearing in higher forms of animal life, is to have circulating blood to carry supplies to the cells and waste products away from them. Most complex animals, including humans, are built on a general plan very much like our unsuccessful imaginary animal, with one important difference. The deep and remote layers of their bodies are served by a system of transport tubes, ducts, or cavities through which fluids can move.

Usually the circulatory fluid contains many vital materials in addition to water – cells, minerals, dissolved food, oxygen, special proteins and so on. When it does, we call it blood.

Life-giving liquid

The simplest animals to have blood are some of the roundworms, but their blood is not confined to blood vessels like ours. Instead, they have blood surrounding and bathing their intestines and filling all the spaces between their other primitive organs. Beginning with the segmented worms, a large group which includes the earthworms and many marine worms, blood circulates in a closed system of tubes lying between the intestine and throughout the body tissues.

Although these segmented worms have a very simple circulatory system and humans have a very complex one, both have the same chemical, haemoglobin, to carry oxygen to the cells that need it. Haemoglobin combines with oxygen and so it makes blood an

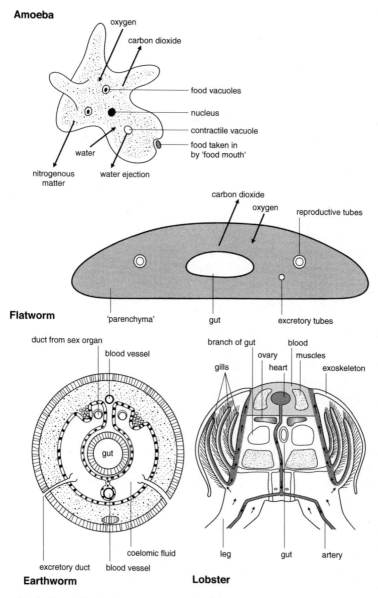

Figure 1.16   Exchange in simple animals

efficient oxygen-carrier. The wonder of haemoglobin lies in its capacity to pick up oxygen when it is abundant and release it precisely where it is needed in the body, without losing any on its way around the body. This is because haemoglobin readily combines with oxygen in body parts where the oxygen concentration, or partial pressure, is high (such as in the lungs). Where the partial pressure of oxygen is low (such as in respiring tissues), the haemoglobin gives up its oxygen.

The haemoglobin molecule is made up of subunits, each of which has a protein chain, called haem. In animals with backbones (the vertebrates), haem has an atom of iron in the middle of the molecule, but in shellfish, the iron is replaced by copper, so their blood is bluish rather than red. One group of marine animals called the sea squirts actually use the trace element, vanadium in their type of blood pigment, and how they manage to extract enough of this from the sea for their requirements remains a mystery. The blood in your body contains about 150g of haemoglobin. It is renewed continually; the worn out haemoglobin is broken down in the liver and spleen and its iron is recycled. New haemoglobin is made in the marrow of some of your bones. It needs iron and a trace of Vitamin B12 which contains cobalt. If either of these is lacking, the person will become anaemic. Indeed, before it was learned how to synthesise Vitamin B12, treatment of pernicious anaemia consisted of eating raw liver, which is its richest source. Cooking the liver will break down Vitamin B12. The American physician, George Minot (1885–1950), shared the Nobel Prize in Physiology and Medicine in 1934 for this discovery.

In vertebrates, haemoglobin is confined to disc-shaped red blood cells, with the claim to fame of being the only living cells in humans which do not have nuclei. As the nucleus controls the cell, the red cells can only have a limited life. In the case of humans, it is about 108 days. They are so small that about five and a half million could fit into a cubic millimetre of human blood – if stretched out, end to end, they would reach around the Earth four times!

In 1908, the Russian biologist, Elie Metchnikoff (1845–1916), was jointly awarded the Nobel Prize in Physiology and Medicine for being the first to detect and describe certain white blood cells, called phagocytes. They were found to be capable of consuming other cells and similarly preying on invading bacteria. This discovery was an important milestone in immunology. For every white blood cell in your body, there are about 600 red blood cells.

Around the same time as the discovery of white cells, the Canadian physician and celebrated lecturer, William Osler (1849–1919), was describing the tiny cell-like platelets found in blood. They help the blood to clot, so that a wound does not bleed indefinitely.

All of the blood cells are carried in plasma, the liquid part of blood. Plasma contains about 7.5% of dissolved proteins of several kinds, and about 0.5% of sodium chloride, which gives blood its salty taste. There are also all sorts of other molecules on their way to and from the body's various tissues. They include, glucose, amino acids, fats, vitamins and minerals as nutrients, together with urea and carbon dioxide as waste products. It is a fact that an analysis of your blood will give information on what has been going on in your body for several days or more. Particular molecules, called hormones (chemical messengers) are secreted from certain glands into your blood and carried to distant tissues to control your metabolism. Very delicate mechanisms keep the supplies of salts and glucose and other important chemicals at precise levels. They also make sure that the blood never becomes too acid or too alkaline, and that it is always ready to carry oxygen and carbon dioxide. Blood has many functions, particularly that of keeping conditions exactly right for the cells that make up our bodies.

The Valentine connection

'I want you to design a pump', the Director said to the Production Engineer.

'That's what I'm here for', the Production Engineer replied confidently.

'Now listen to me', the Director went on, 'the pump must keep eight pints of fluid on continuous circulation at a temperature of 37ºC. It must work against constantly changing resistance and must adapt itself in an instant, sometimes pumping eight pints per minute, sometimes fifty.'

'What do you mean by in an instant?'

'Within a tenth of a second.'

The Production Engineer frowned and gave a whistle through his pursed lips.

'The pump must weigh no more that 350g. It must work at the height of Mount Everest, or in the Sahara Desert, or at the North Pole. It must work by day and night and at 130 strokes per minute – then in a couple of minutes, drop to 70. It will be controlled by a self-regulating, electrical mechanism.'

'You mean automation?' the Production Engineer asked. 'We are quite used to that.'

'The pump must also respond to human control – which at any time may counteract the automation. The controller of the pump may be asleep, or drunk – it must make no difference.'

'It's going to be hard to design a pump like that.'

'Oh, and another thing. The pump will have four chambers with four valves, and has to drive the fluid in opposite directions at the same time.'

'What about running repairs?'

'Servicing will have to be done without the pump losing a single stroke. The pump must never stop. Oh, just one other minor point. The pump must be able to work for 100-plus years.'

At this point, the Production Engineer let his pencil fall. 'It can't be done sir! A pump like that simply could not be designed.'

Then the Production Engineer saw a smile of triumph on the Director's face.

'Oh, I see. It is an imaginary pump; it's your dream – for the future.'

'No, it is a real pump. You have one under your shirt. The finest pump that was ever developed. It grew on its own and what's more, mine has been going for half a century – with automatic servicing. There has never been anything like the human heart – it's nothing but a pump with tubes attached. Four chambers, two circulations, all kept up with perfect automation. And it even works when we are not thinking. It can respond to any requirement – heat and cold – bad temper, running a mile. It is truly a lesson in design.'

'But the real point is', here the Director paused, 'the heart does go wrong. The material wears out before it should, those tubes become rough inside and are silted up. The valves get glued up. The automation goes wrong and we simply don't know enough about how the pump works, to put it right.'

Even in the twenty-first century we do not pretend to know all of the reasons for heart failure. However, we have come a long way since people began to understand the circulation of blood. Although, attributed to William Harvey (1578–1657), the first sensible ideas on the mechanism of blood circulation were probably developed previously by Andrea Cesalpino (1519–1603). He envisaged a blood circulatory system at the time when he was physician to Pope Clement VIII.

Undoubtedly, one of the greatest English physicians in the history of medicine, William Harvey is regarded as one of the founders of anatomy and a major contributor to medical theory. Born in Folkstone, Harvey studied at Cambridge before studying at the University of Padua. On returning to his homeland, Harvey was appointed as a physician at St Bartholomew's hospital and lecturer to the Royal College of Physicians. In the years between 1612 and 1628, he undertook a remarkably detailed study of the cardio-vascular system, culminating in his major work *De motu cordis* in which he announced the theory, heretical at the time, of blood circulation.

Harvey's conclusions were classic in their simplicity and lacked only one element to complete the account of circulation. He did not demonstrate the existence of capillaries, the observation of which required a more sophisticated microscope than Harvey had. Shortly after Harvey's death, the Italian anatomist, Marcello Malpighi (1628–1694) was able to complete the story of circulation by observing blood cells trickling, almost one at a time through capillaries. He published his observations in 1661 in his classic study of the lungs, *De Pulmonibus*.

So, in the ensuing 300 years, the picture we have built up of the heart and circulation is as follows (see fig. 1.17, p.60).

Around forty million times a year, on average, our hearts beat to drive blood through arteries which become narrower and narrower until they can only be seen with a high-powered microscope. These blood vessels are then known as capillaries which, in turn, gradually become wider and wider to form veins which return blood to the heart. Essential materials must be delivered to cells, and waste products must be removed. The billions of cells in our bodies must be fed, watered, and cleaned or else they will die. Our hearts can never stop their ceaseless, controlled pumping.

The simplest animals that have blood and blood vessels are heartless. The blood is circulated by body movements. More complex animals – segmented worms, insects, and shellfish – have

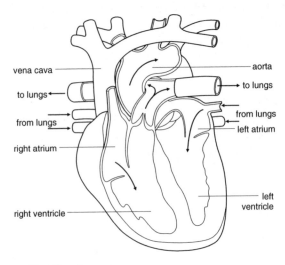

Figure 1.17 The heart

muscular thickenings at one or more places along the blood vessels, which pump the blood by contracting rhythmically. From these primitive pumps the more complex four-chambered hearts of mammals have evolved.

There are two principal blood circuits in mammals, one through the lungs, and one through the rest of the body. Each has its own arteries and veins, joined by tiny capillaries. The networks of capillaries in active tissues are more dense than in less active tissues. When a tissue like a muscle is not being used, the muscular walls of the small arteries supplying the tissue contract, so that less blood flows through them.

In a baby before birth, blood does not circulate through the lungs (except to help them develop). It is pumped through an artery in the long umbilical cord to a system of capillaries in a special organ called the placenta. This is the vital point of union between the developing baby and its mother. The capillaries of the placenta are in close contact with the mother's blood in the wall of the womb, or uterus, in which the baby develops. Oxygen and dissolved food diffuse from the mother's blood to the capillaries of the baby. Carbon dioxide and urea diffuse in the opposite direction and are disposed of by the mother's lungs or kidneys.

The baby's haemoglobin combines even more readily with oxygen than does the haemoglobin of an adult. Blood returning along the cord goes first to the baby's liver, to deposit food, then to the heart and brain to carry oxygen to these organs.

After the baby is born and begins to breathe, blood begins to circulate through its lungs. The placenta then separates from the uterus and is expelled. It can no longer supply food and oxygen, so the umbilical cord is cut.

Waste disposal

Life is a constant race with death. While cells live, they must constantly maintain a balance between the processes that produce order and structure, and those that destroy. The activities that feed cells must be balanced by those that carry away wastes and by-products that cells cannot use. Some of the chemicals produced in excess are carbon dioxide, water, salts, urea, ammonia, and uric acid. The important function of the waste disposal system or excretory system, is to get rid of poisonous wastes from animals.

In mammals, like ourselves and in other complex animals with well developed blood systems, the substances that must be supplied to and removed from cells, are carried in the blood. The excretory system tends to keep the composition of the blood constant as part of homeostasis (the regulation of the internal environment). Excretion is the elimination of chemical waste products from chemical reactions which have taken place in the body and is essentially different from elimination of faeces. The latter is the removal of material from the digestive system. This waste material is simply undigested fibre (cellulose) and bacteria. The material has never been part of any chemical reactions in the body.

Plants do not have any organised excretory system. Carbon dioxide and oxygen diffuse through the pores of the leaves and stems. Carbon dioxide is a waste product of respiration and diffuses out during the night because, at that time, it cannot be used for photosynthesis. Oxygen is a product of photosynthesis during the day, and if it is in excess of the plants requirements, it is passed out as an excretory product. Transpired water exits through the pores of the leaves and stems by evaporation.

Animals are different in that they cannot store protein which is in excess of their needs. Protein is broken down to amino acids, and if these are not required for growth or repair, they are changed into chemicals which can be excreted.

In simple animals that live in water, like marine worms, the amino acids are changed into ammonia which then passes out into the surroundings. Their excretory systems consist of tubes, the nephridia (little kidneys) surrounded by blood capillaries. Waste molecules diffuse out of the blood into the excretory tubes.

Marine fish swallow sea water all the time, but with it they have to take salts they do not want; these are excreted through special cells in their gills. Sea birds, like the albatross and storm petrels get their water from the sea. These also have special glands at the base of their beaks to pump out the salts.

Urea is very soluble and harmless unless in very high concentration. It can be carried away by less water than is needed to carry the much more toxic ammonia away. Urea is the main form in which waste, containing nitrogen, is excreted in mammals. By the pressure of the heart beat, arterial blood is forced through filters in the kidneys. These are the Malpighian bodies – about a million in each kidney. They are named after their discoverer, Marcello Malpighi (1628–1694), an Italian anatomist who was a pioneer of microscopic examination. Small molecules are forced through these filters into a series of tiny tubules. Proteins, however, are usually too big and remain in the blood. As the filtered material passes along the tubules, the cells take back useful substances, like glucose, amino acids, and any water and minerals that may be needed to keep the concentration of body fluids constant. Urea and other waste molecules remain in the urine which passes into the bladder.

In animals that lay eggs on land, like insects, reptiles and birds, the waste products, ammonia and carbon dioxide, are changed into uric acid. In the egg, the harmless, insoluble uric acid accumulates as the waste product of the developing embryo and is discarded on hatching. In the adults, dissolved waste is concentrated into solid uric acid, passed into the end of the digestive system and expelled with the faeces.

How animals get their breath

In everyday terms, we tend to use the work 'breathing' to mean 'respiring'. Biologically speaking, this is not strictly correct. The words mean two different things although you can't have one without the other. Respiration is the chemical process of releasing energy from glucose in all living cells. On the other hand, breathing is a physical process of gas exchange relying on diffusion at a surface which is modified for such a task.

Breathing surfaces vary from simple coverings like the outer layer of an earthworm, to gills, or to the most complex system of all – the millions of tiny air sacs in our lungs and in those of other mammals.

In all cases, however, in order to be efficient, the surfaces must satisfy the following conditions. They must have a very:

- large surface area in relation to the bulk of the animal
- rich blood supply
- thin surface – usually one cell in thickness
- moist surface because gases diffuse in solution

In addition to these features, all active animals must have a good means of renewing the air at the breathing surface – ours is like a bellows. Fish use movements of the mouth to force water over the gills.

The cells of insects are not supplied with oxygen by means of blood, but by a network of fine air tubes that branch from openings along the sides of the body. Movement of the body muscles help to move air along the tubes, but oxygen is replaced and carbon dioxide removed by means of diffusion. The system is very efficient for animals up to the size of the largest insect in the world; but it would not supply enough oxygen to tissues deep inside larger animals like an elephant or a whale.

Gills and lungs contain networks of thin-walled blood capillaries located close to the surface of these organs. As water or air flows over the surface of the gills or lungs, oxygen diffuses into the blood. In humans, the surface exposed to air in the lungs is 25 to 50 times larger than the whole surface of the body. Some fish and frogs have moist skins and get oxygen through both their skin and their gills or lungs.

In mammals, ventilation of the breathing system takes place by movements of the rib cage and with the help of a muscular partition between the abdomen and thorax, the diaphragm. The combination of actions of these structures will alter the volume and the pressure inside the lungs causing breathing in and out. The need for more oxygen is satisfied by breathing faster and deeper, which inflates the lungs more fully and exposes more surface area for diffusion.

The rate of breathing is controlled by a part of the brain which reacts to the amount of carbon dioxide in the blood at any time. When the muscles start to use glucose faster, the carbon dioxide in the blood rises above normal. The breathing centre in the brain sends an urgent message along the nerves to the muscles between the ribs and to the muscles of the diaphragm, telling them to work harder until the extra carbon dioxide level in the blood is reduced to normal once again. On the other hand, if the blood contains too little carbon dioxide, the breathing centre makes the respiratory muscles work more slowly. Only a small amount of oxygen is dissolved in the liquid portion of the blood. Most of the oxygen diffuses directly into red blood cells where it combines with haemoglobin, which later releases it to cells in other parts of the body. Carbon dioxide also is not merely dissolved in the plasma of the blood. It enters into chemical combinations as well. It actually combines with water to make carbonic acid and then, after a series of complex reactions, becomes sodium hydrogen carbonate. When blood reaches the lungs, this is reversed; carbon dioxide and water are released to be breathed out.

Even when you are fast asleep, carbon dioxide and oxygen must be carried in your blood and exchanged in your lungs. The heart and diaphragm never stop working. Even though your arms and legs may be perfectly still, your liver, kidneys, brain and other internal organs are active.

The rate at which a resting person uses oxygen and gives off carbon dioxide after a night's sleep and no breakfast is called the basal metabolic rate (BMR). This is a measure of all the life processes going on in the body when they are subject to no stress or strain. It increases with the surface area of the body; hence ten tiny babies use much more oxygen for basic cell activities than one large adult. BMR tests are important in medical diagnosis. Once an average for age, sex and size is known, a patient's BMR can be compared with it.

Armour and support

Land living, many-celled animals and plants are restricted in size unless they have a means of rigid support. Without such a structure, the organism would collapse under its own weight. In animals, in addition to serving as a support for soft tissues, such structures help in the process of movement, and in many cases, protect organs. For example, our skulls encase our brains like a protective hard hat and our ribs form a cage to protect our heart and lungs. Vertebrates (animals with backbones) have bony skeletons hidden on the inside of their bodies. These are called endoskeletons (from *endo* = inside). Many invertebrates (animals without backbones) have their skeletons on the outside of the body armour plating. We commonly call these exoskeletons (from *exo* = outside) shells. The biologically inaccurate term, shellfish, is often used to cover molluscs like clams and oysters as well as crustaceans like crabs and lobsters. The term is inaccurate because these animals are not true fish and the exoskeletons of molluscs and crustaceans are built in an entirely different way.

Even some one-celled animals and plants have hardened exoskeletons outside their cell membranes. Astronomical numbers of these lived and died millions of years ago in ancient oceans and now make up the great white chalk cliffs of Dover on the south coast of England. These deposits of chalk are cretaceous rocks (from the Latin *creta* = chalk). Imagine a continuous snowstorm taking place in the sea over millions of years. The flakes are the skeletons of single-celled marine animals called *Foraminifera*. Oceans were alive with them, as indeed they are today in many parts of the world. They were so numerous that the floors of the oceans became covered with layers of them, which became compressed into chalk. The deposits have been lifted to their present positions by movements of the Earth and they show that, in times gone by, great seas existed where they no longer occur today. The stone of many of the most historically important buildings in the world consists of these fossil microbial shells. Indeed, the Egyptian pyramids were made from this material. Other single-celled creatures have hard, rod-like bodies on the inside of the cell which seem to function as supporting structures. Clams, oysters and snails have exoskeletons made mainly of lime (calcium carbonate). The insects and crustaceans are covered by jointed armour plating made up of a hard, protein-based substance called chitin.

The external skeleton that makes up a large part of the weight of insects and crustaceans gets disproportionately heavier as these animals grow. Thus it limits the size of the animal and also has to be discarded when the animal grows – in the case of crabs and lobsters. At such times, these animals are very prone to attack by predators and they must hide until a new hardened skeleton has grown. Once insects have developed adult features, they do not grow any more.

An internal skeleton, on the other hand, overcomes the problem of restriction on size to some extent because it grows with the rest of the animal. All really large animals like elephants and whales have internal skeletons. Indeed, the largest invertebrate, the giant squid, is an exception to the normal plan of invertebrate skeletons, and has an internal supporting skeleton.

When bony fishes evolved, the basic plan of all vertebrate skeletons first appeared:

- ■ a skull which protects the brain:
- ■ a jointed back bone made of separate bones called vertebrae
- ■ paired limbs

When the fish-like ancestors of modern land animals began to take up life on the shores, fins evolved into jointed limbs and became important as locomotory organs for movement on land. The bony shoulder and hip girdles became larger, as did the limb muscles that are attached to them. The tail became less and less important for movement until it was lost altogether in apes and humans.

The remains of the tail is still found at the base of our backbones (the coccyx). Indeed, human embryos have tails in the early stages of their development showing traces of our ancient ancestors.

Meat is muscle

A skeleton without muscles is like a chassis of a car without a motor. If you turn a somersault, you use hundreds of muscles all over your body. Even raising your arm requires the work of several muscles. When a muscle works, it contracts. The resting muscle cell is long and thin, but when in action, it becomes short and fat.

Muscle cells are grouped in bands. If the ends of a band are attached, by tendons, to different bones and across a joint, a contraction of the band brings the bones nearer together. When you want to display your muscles, you shorten the biceps muscle in your upper arm and you bend your elbow. When you consciously release the tension, gravity causes your arms to drop again. Not all muscles are attached to bones. The muscles of the heart, arteries and intestine do not move our bodies, but perform other essential tasks such as forcing liquids along.

Muscles range in variety and complexity. Jellyfish have simple muscular fibres on their inner sides which contract and relax so they can pulsate through the water. In more complex animals there are three types of muscles. Those capable of quick contraction have cells which appear under the microscope to have crosswise stripes. They are called the voluntary, or striated muscles, with a nerve supply placing them under the control of conscious effort.

The muscles which move internal organs and blood vessels are involuntary muscles, with unstriated or smooth cells and a different sort of nerve supply. The heart's muscles belong to neither of those groups. The cells are striated both crossways and longways, and their contractions originate in the muscle itself, though they can be controlled by nerve messages.

The most active and largest striated muscle cells known are attached to the wings of insects. Some can cause the wings to beat at 300 times per second! Birds also have wonderfully active muscles; the hearts of the smallest birds can beat hundreds of times per minute compared to the average human whose resting heart beat is about 80 beats per minute.

An active muscle needs oxygen and glucose as fuel to supply energy, and a steady supply of protein as building material to keep it in a good state of repair. Activity produces carbon dioxide that has to be removed. So muscles need a good blood supply. Blood is supplied to the muscles, according to need, by capillaries. These tiny vessels open when the muscle is busy, and close when it is resting.

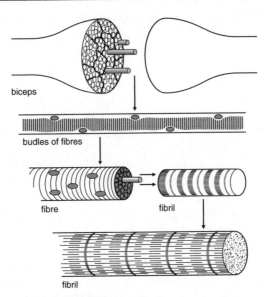

biceps

budles of fibres

fibre   fibril

fibril

Figure 1.18   Muscle structure

The contraction of a muscle occurs when it receives a message along a nerve. In humans, these signals travel from the brain at more than 200 miles per hour. Muscle cells have an ingenious chemical arrangement by which energy is stored ready for use. The energy source for muscle contraction is our old friend, the power-packing phosphate molecule, ATP (see p.32). In muscle cells, ATP receives its energy from glycogen, a carbohydrate compound made of many glucose units – the animal equivalent of the starch stored in plants. As a first step in muscle action, glycogen is split into glucose molecules by enzymes. From that point on, the reactions are very similar to the ATP building reaction in respiration (see p.34). Here again the first steps are accomplished without oxygen and, if necessary, muscles can keep on contracting until all the glycogen is used up.

However, without oxygen, the chemical changes in glucose produce lactic acid. With too much of this, muscles become tired and eventually cannot respond. Most of us have experienced the pain in our muscles if we keep them contracted for too long. See how long you can hold your arm out straight before feeling the pain of accumulation of lactic acid. Some of the lactic acid passes into the blood and warns the breathing centre of the brain that more oxygen is needed. We now breathe faster and deeper. With the available oxygen, some more lactic acid is oxidised to provide energy, which is used to rebuild glycogen from glucose units. In a resting muscle or one doing moderate work, oxygen is supplied as fast as it is needed. In violent exercise, the muscle goes into 'oxygen debt'. Lactic acid accumulates, only to be removed when the oxygen supply allows. Athletes who exert themselves to their limits have a large oxygen debt in their muscles which is repaid with oxygen obtained in the deep panting breaths of exhaustion.

Wiring our computer

Messages from our brains control all our actions as a result of a mind-boggling computer system which, to date, has not really been simulated with all the information technology available to us. Clues to the evolutionary development of our brain and nervous systems can be obtained from a comparative study of simple and complex animals. One of the simplest nervous systems is found in *Hydra*, a member of the jellyfish group of animals. It consists of a network of interconnecting nerve cells spread throughout the body between the two layers which make up its bulk. There is no central control. Thus, if one part of the animal is stimulated, the entire body responds rather than the particular region which has been stimulated.

Flatworms have a somewhat more specialised system. Here, the nerve cells are arranged in two longitudinal nerves connected by transverse nerves, producing a 'ladder-like effect'. They also have collections of nerve cells at the head end forming swellings called ganglia. These are the forerunners of the brain of more complex animals.

A more complicated nervous system is found in the earthworm. Here we see a single longitudinal nerve running along the underside (ventral) just below the main blood vessel and the digestive system. The primitive brain consists of two cerebral ganglia about the size of pinheads. At intervals along the main nerve, paired nerve branches emerge and supply the various organs of the body. Vertebrates have brains which are built on a common plan showing evidence of a common ancestry. In the primitive vertebrates like fish, amphibia and reptiles, three parts can be clearly seen; hindbrain, handling nerve messages from the muscles, skin and ears; a midbrain, particularly concerned with the sense of sight; and a forebrain, mainly concerned with the sense of smell. In higher vertebrates, like birds and mammals, there is a great development

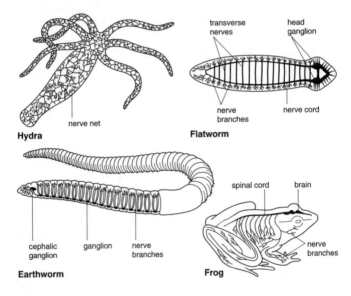

Figure 1.19   Nervous systems

of a part of the forebrain called the cerebrum, which contains cells that perform the functions of memory, learning and other higher mental powers.

Movement and balance can be automatic and all vertebrates have to master them for survival. Again, there is a basic plan of action which is common to all but the details of mechanisms which vary according to the mode of life of the individuals. For example, flying, swimming and running all require their own adaptations of a basic plan of co-ordination. However, they all begin with special sensory organs in muscles which report the muscles' state of contraction to the brain. This information is needed before the brain can 'order' that muscle to act. When a message, or impulse, passes along a nerve, it does so in a way that is quite similar to an electric current passing along a wire. Your nerves are each like little charged batteries which convert chemical energy into electrical energy. Where a nerve meets a muscle it stimulates it by setting free a chemical which 'switches on' the muscle contraction. As soon as this chemical has done its job, an enzyme breaks it down so that the muscle can relax again.

Sometimes, even humans with the ultimate computer in their skulls are not aware of certain actions that are performed automatically. Our internal parts may have to be active for 100 years or more without stopping. If we had to remember when, and how fast to breathe, or when our hearts should speed up, most of us would forget sometimes and live shorter lives! There is a combination of automatic, chemical and electrical control over these and other activities which allows them to go on night and day at a rate which is optimum to our requirements.

There are also, fast, unconscious, automatic responses to stimuli which we carry out to protect ourselves from danger. These are the reflex actions – truly thoughtless actions. Blinking, sneezing, contracting our pupils in bright light, and withdrawing our finger from a sharp object or a flame, all go on without us having to think about them. Basically, a stimulus (any change in the surroundings) causes a sense organ to respond and cause an electrical impulse to pass along a sensory nerve to our central nervous system (brain or spinal cord). The impulse is now relayed to a motor nerve which then passes to a muscle or a gland to cause a response. Sometimes we become aware of the process and can even control it because the brain is informed and we can sense pain. At other times, we have no control at all over the action and, indeed, do not even know it is happening, for example when our pupils in our eyes contract in bright light.

Our detective agency

Humans seem to have an obsession with intelligence and that great, grey blancmange-like blob from which it stems – the brain. However, a certain irony arises here because we don't really know what intelligence is. Scientists have been wrestling with an all-purpose definition for many years. The result has been sharp

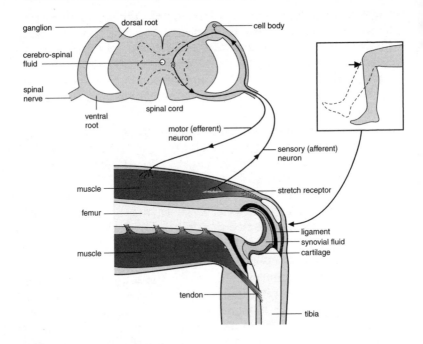

Figure 1.20 Thoughtless actions

disagreement and a complete lack of consensus between psychologists, philosophers and neurologists. When biologists try to apply the word to species other than humans, 'intelligent' discussion degenerates into a shambles. Yet we believe that intelligence, whatever it is, is good – at least for us. We believe that it somehow resides in the brain.

You may wonder why the brain is located in the head. There have been exceptions, almost like evolutionary experiments which have led nowhere in particular. Some of the extinct, massive, herbivorous dinosaurs were such freaks of nature. They had a second 'brain' at the base of their tails to help to orchestrate their immense bodies as they browsed in prehistoric forests. The evolutionary reason for command posts being in the front of animals is to permit quick analysis of the environment in which the animal is advancing.

These nerve centres, as headquarters, would in all probability have come to be associated with the specialised receptors we refer to as the senses. The sense organs responsible for sight, sound, smell and taste are most commonly located in the head region. If the brain and these special receptors were located at the posterior end, the animal might find itself in an inhospitable environment by the time it realised its predicament.

Besides those sense organs which tend to be based in the head, there is one which surrounds all of us. It is the skin. Ordinarily we do not pay much attention to skin except to keep it clean. Skin, however is more than a bag for bones and organs. It protects, helps to regulate your temperature, and senses certain changes in the environment. It has sensory nerve endings. In the deeper layers there are nerve endings for detecting pressure, and nearer the surface there are detectors for temperature changes.

The tongue, like fingertips, is well supplied with nerves that respond to pressure; and it has a special set of nerve endings in the taste buds. They react to dissolved chemicals and report to the brain four or five kinds of flavours – sweet, sour, salt, bitter, and possibly metallic. No doubt you have experienced other flavours, but in fact, all other tastes are smells detected by a special surface at the back of the nose. When you have a cold, this surface may cease to work properly and food seems tasteless. The reason is because the accumulation of mucus and cell debris on the surface prevents chemicals reaching the sense cells. Humans have a much smaller area of smell-detecting surface than cats or dogs.

Another sense that provides necessary information about the environment is that of hearing. The eardrum vibrates because it is struck with vibrations in the air which are eventually interpreted by the brain as sound. The eardrum passes the vibrations on to three very small bones, and then to a membrane with short, tightly stretched fibres at one end, grading into longer, looser fibres at the other. Nerve endings connected to these fibres report to the brain, which translates the messages as a series of sounds.

Also, deep in the apparatus that we know as the 'inner' ear, there are three semi-circular canals. They are made of loops set in three planes at right angles to each other, filled with fluid into which project tiny hair-like structures which can detect changes in the fluid of the canals. Nerves in these sensitive 'hairs' inform the brain of the position of the head and help you keep your balance. When you spin around and then stop suddenly, the canal fluid goes on moving for a short time, telling your brain that you are still turning around. This makes you feel dizzy. Similar systems for balancing and keeping animals the 'right way up' are seen in creatures as primitive as jellyfish, shrimps, and lobsters.

The most complex of the senses is that of sight. In humans and other vertebrates, the optic nerve between the eyes and the brain may have more than a half a million nerve fibres. They encode into nervous impulses, the complicated information transmitted when light energy is converted into chemical energy and then into electrical energy. The rays of light are focussed by various parts of the eye, including the lens, and meet on the light-sensitive layer called the retina. Here, special parts of nerve cells called rods and cones begin the process of energy conversion.

Insects and crustaceans have eyes which are made up of thousands of individual units, acting independently. The result is an image which is like a mosaic or a newspaper image as seen with a magnifying glass. Vertebrate eyes and those of invertebrates have evolved separately. Even though the end product, sight, is the same, it has been achieved by two dissimilar paths of evolution. The combination of the actions of our sense organs and nervous systems means that animals function as whole organisms, rather than a collection of millions of independent cells.

Chemical controllers

In the previous pages, there are accounts of many activities of living things; what food they need, how it is digested, absorbed, built into cell substances or used to release energy; how they respire and excrete; and how they move, react and adjust to their environments. All these activities go on at one time. Somehow, all the many activities of organisms are co-ordinated.

In the words of Claude Bernard (1813–1878), 'they are directed by an unseen guide'. Bernard, the son of a grape grower, became one of the greatest French scientists. Besides having great literary talent, he graduated in medicine in Paris and became a brilliant experimental biologist. The most outstanding of his general concepts was that of the '*milieu interieur*' or internal environment being automatically controlled by chemicals. Today, we call the idea homeostasis or organic equilibrium. By suggesting the existence of internal secretions, he became the father of endocrinology or the study of hormones.

Bernard's 'invisible guides', with which we have become so familiar, were called hormones from the Greek word meaning 'rouse to activity'. The term was coined by two English physiologists, William Maddock Bayliss and Ernest Henry Starling, in 1902. Many hormones are protein-based and all are secreted directly into the bloodstream by organs called endocrine glands.

Incidentally, the word 'gland' is derived from the Greek word for acorn. Strange to us perhaps, but the name was given to any 'lump' of tissue found in the human body by anatomists of the ancient world. Plants, too have chemical regulators which control growth and are called plant hormones.

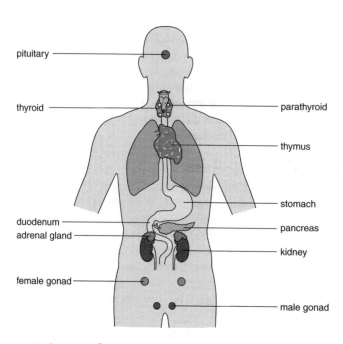

Figure 1.21   Endocrine glands

Plant hormones, however, are made in the dividing cells of the growing tips of the plants' stems and are passed to the cells lower down. They stimulate cell growth in stems but seem to have the opposite effect in roots.

In higher vertebrates, the basal metabolic rate – the rate at which oxygen is used by a resting individual – is controlled by a hormone, thyroxine, entering the blood from the thyroid gland in the neck. Normally, thyroid activity is adjusted perfectly to the needs of the body. If disease causes the human thyroid to make too little hormone, the 'fires' burn slowly; the result is lethargy, often accompanied by obesity. A baby with too little or too much thyroid hormone, will not develop normally. Lack of iodine in the diet is one cause of thyroid hormone deficiency because iodine makes up part of this hormone's molecule.

As long ago as 1760, the Italian, Giuseppe Flaiani, physician to Pope Pius VII, described the condition known as *exophthalmic goitre*. In 1840, the German physician, Karl von Basedow confirmed Flaiani's work and noted that goitres grew in people who lacked iodine in their diet. The condition is a large swelling in the neck where the thyroid gland produces masses of cells in an attempt to make up for the deficiency. In regions where traces of iodine are found in drinking water or where it is added to table salt, the condition is now very rare. It can be treated by surgical removal. Thyroxine was finally isolated and named in 1915 by the American biochemist, Edward Calvin Kendall.

The little adrenal glands which lie just above the kidneys make a hormone called adrenaline. This is poured into the blood when an animal is stressed and prepares it to run or fight. The blood pressure rises, the heart beats faster and more strongly. Blood vessels in the muscles dilate to carry more blood to them and the tubes in the lungs leading to the air sacs dilate so that more air can reach them. The adrenals also make hormones called cortisones which control the salt balance of the body and the way that glucose is metabolised. The pancreas, which pours digestive enzymes into the digestive system also has special cells which secrete the hormone, insulin, into the bloodstream.

Insulin helps to regulate the concentration of blood sugar, glucose. In 1923, the Canadian physician, Sir Frederick C. Banting, shared the Nobel prize in Physiology and Medicine with the Scottish physician, John James Macleod for isolating insulin. Banting, together with another Canadian physiologist, Professor Charles Herbert Best, revolutionised the treatment of diabetes because sufferers are able to inject themselves with controlled doses of pure insulin. In 1934, Banting received a knighthood from King George V for his outstanding contribution to medicine. As a result of his work, how many lives have been saved on a world-wide scale? Ironically and tragically, his own life was wasted when he was killed in an air crash in 1941.

People with *diabetes mellitus*, either do not produce any insulin or cannot produce enough for their needs. Instead of being able to change excess glucose in their blood to glycogen which is stored in the liver, the glucose builds up in concentration in the blood. As a result, glucose is excreted in the urine, leading to loss of essential fuel for the body's supply of energy, and possibly ending in coma and death. This symptom was the reason for giving the name *mellitus*, from the Greek word for honey – referring to the sweet taste of urine from *diabetes mellitus* sufferers. This observation could have been made by an heroic or eccentric physician! Legend has it, however, that the observation was rather less direct than this and was made by someone noting the increase in flies gathering around urine from a diabetic person. In 1815, the French chemist Michel Eugene Chevreul positively identified glucose in the urine of diabetics. Today, diabetics are treated by daily injections of insulin, either extracted from the pancreas of an animal or produced by genetically modified bacteria (see p.109).

The problem not only affects the entire energy economy of the body, but all the reactions that require ATP (see p.32). Moreover, because excreting the load of sugar presents the kidneys with a task similar to excreting syrup, great quantities of water are required to dilute the urine. This, in turn, upsets the entire fluid balance of the body.

Apart from insulin, sex hormones are probably the most talked-about hormones in our society because of their use in contraception, fertility treatment, and in hormone replacement therapy (HRT). The origin of the development of medical treatment by these hormones goes back to 1927 when two German physiologists, Bernard

Zondek and Selmar Ascheim, discovered that extracts of the urine of pregnant women, when injected into female mice, aroused them to sexual frenzy. This test actually led to the first early tests for pregnancy. Within two years, pure samples of sex hormones were isolated and given the name oestrone, from *oestrus*, a term for sexual heat in females. Oestrone is now put in a group of known female sex hormones called oestrogens. In 1931, another German, Adolf Butenandt isolated the first male sex hormone, androgen, meaning 'giving rise to maleness'. Butenandt, together with several other German scientists was forced to reject the Nobel Prize until after the destruction of the Nazi tyranny of World War II.

The ovaries and the testes produce sex hormones as well as sex cells. The hormones govern what are called the secondary sexual characteristics – shape of body, pitch of voice, the distribution of hair in men and women. The ovaries are also active in the control of the menstrual cycle in women. Here, several hormones interact to regulate egg release and the preparation of the womb lining for pregnancy. Such hormones are artificially used for the treatment of infertility and for contraception. Hormone replacement therapy in post-menopausal women depends on the fact that sex hormones can diffuse into the blood system through the skin.

At the base of the brain, just above the back part of the throat lies a small structure, about the size of a pea. It is the pituitary gland and is the master endocrine gland of the body. It is called the 'master gland' because it produces hormones which control all of the other endocrine glands in the body.

2 | THE LIFE MACHINE

No sex please – we are asexual!

Yes, it is possible to go through life without sex. Reproduction may involve only one parent (asexual) or may require two parents (sexual), but in either way, hereditary information passes from the parent to the daughter cells via chromosomes.

When some one-celled creatures reproduce by fission – splitting in two – the parent cell simply disappears. Unlike the method of producing new life with which we are most familiar, there is no requirement for a union of male and female sex cells. In fact, asexual reproduction is widespread among living things.

Both amateur and professional gardeners reproduce plants by the simple and direct method of cutting off a leaf or branch and placing it in water or damp sand until it roots. The process can be speeded up by first dipping the cutting into rooting powder containing plant hormones. Strawberries and many other plants send out runners, or stolons, stems that grow along the ground and produce roots. Bulbs form new bulbs from buds inside themselves. All these are familiar forms of asexual reproduction. However, most of these have best of both worlds by also having methods of sexual reproduction via flowers.

In the animal world, one-celled creatures are not the only ones to increase by asexual means. There are marine worms that break into pieces, each of which grows into a new worm. Flatworms can reproduce by anchoring their tails to the ground while the front end seems to lose interest in its back end and crawls away! Each half eventually regrows the missing parts. *Hydra* can grow little ones on its body as buds. Sea anemones are *Hydra's* colourful cousins and are usually found fastened to rocks by a sort of disc of tissue.

Sometimes they contract, pulling in this disc but leaving behind a ring of tissue. In a few weeks, in favourable conditions, the discarded pieces may grow into new anemones.

Bees and wasps can often lay eggs that have not been fertilised by sperm. Worker bees, which are females that do not complete their sexual development, are hatched from such eggs. Some insects, such as aphids, produce several generations of females without having come in contact with the male.

A single cell can become two cells by fission; but how would a horse or elephant divide to become two animals? Generally, during asexual reproduction, the whole body is involved in fission and this, obviously, is quite impractical for large and complex animals.

If multi-cellular animals are to reproduce asexually, at least some of their cells must remain unspecialised if they are to develop new individuals. You cannot expect cells which have spent time developing into nervous tissue or muscle tissue to suddenly change direction and become whole new individuals. Plant cells, however, are different and can do this very unexpected trick.

Even cells as specialised as those in a leaf or stem can develop in roots and vice versa. This is the basis of cloning using plant tissue cultures and is carried out in lucrative commercial enterprises all over the world. Important crops like oil palms, bananas and rare orchids are grown in this way. Cloning animals has become an emotive scientific concept since the 1990s and its future depends on attitudes and motivation of politicians, scientists, and an informed public. There is more on this in the section on gene technology (see p.105).

The principles of sex

Although there are probably more variations of sexual reproduction than there are of asexual reproduction, the basic pattern is always the same. Two cells unite to form one cell, the fertilised egg or zygote. From this single cell the new individual develops.

The formation of new individuals by the union of two cells, the female egg and male sperm, eliminates the need for the adaptability of all cells and permits the high degree of specialisation needed in an elephant or a human.

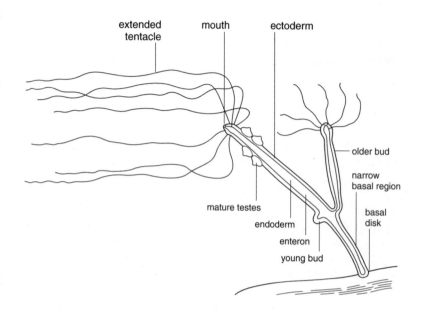

Figure 2.1 *Hydra* **budding**

When organisms reproduce sexually, the only cells that need to retain the capacity to develop into new organisms are the sex cells, eggs and sperm; and their reproduction is limited to special organs in the body, the gonads – ovaries and testes.

Sexual reproduction has various advantages for animals and plants. Seeds which contain the plant embryos are usually produced in vast numbers. They can remain dormant, sometimes for several years, until conditions are favourable for growth. Plants that are limited to asexual reproduction may often be wiped out of an area following several seasons of adverse conditions. By far the most important advantage of sexual reproduction, however, is the certainty of variation among the offspring. This is because the genes of both parents will mix. Variation makes adaptation possible and this, eventually leads to evolution by natural selection. (see p.138).

For or against sex

We must keep in mind that evolution rewards only reproductive output. The genes of individuals who do not reproduce maximally will be replaced by the genes of those who do in the relentless arithmetic of natural selection.

Reproduction, of course, involves passing along replicas of our genes into the next generation. Thus, logically, if we wish to maximise our genes in the next generation, it would be best to produce genetic replicas of ourselves. That is, to pass all of our genes into each of our offspring. Yet many organisms dilute their genes by mixing them with those of a partner through sexual reproduction. Why dilute them in this way? The question of how sex arose has produced a great deal of debate in scientific circles.

Apart from the question of gene dilution, there are a number of other problems associated with sexual reproduction. For example, it requires an enormous number of sex cells although only a few will complete the process of fertilisation – a great waste. Also, for some species, there is a real chance that two individuals, or their sex cells, might not be able to find each other because they are so sparsely distributed. For example, a plant that releases pollen into the air must not be too far from its nearest neighbour or its effort will be wasted. In fact, the probability of breeding failure for members of some species is very high. So if you can reproduce all by yourself, you not only ensure that your offspring will have all of your genes, but you eliminate the risks and waste associated with sexual reproduction.

Why, then, have sex? Your first response to this question notwithstanding, one advantage of sexual reproduction will be apparent if you understand the process by which sex cells are formed (see p.96). The process, called meiosis, yields sex cells that carry different kinds of genes, and normally, there is no way of predicting which of the sex cells will enter into fertilisation. As a result, any individual's offspring are likely to be highly variable. This variation is critical to evolution in an ever-changing planet where all organisms are subjected to unpredictable selective pressures of their environment.

With a genetically variable population, there is an increased likelihood of survival for some members of the population if environmental conditions change. Sexual reproduction also allows for the union of advantageous genes. Suppose for example, that two individuals bearing two newly-mutated 'superior' genes should mate. Those offspring bearing both mutations would be at an advantage. Another advantage of sex is that recessive genes (see p.92) can lie dormant in a population. They cannot be expressed if they are masked by dominant genes and so cannot be subjected to natural selection. These genes can be a reservoir of potential variation should the alternative dominant gene suddenly become a disadvantage.

In populations that reproduce without sex, a new 'good' gene that arises in a population can quickly become fixed (reach 100%). This is fine for the gene, in a sense, but such an event can cause problems. Suppose an asexual population of pond dwellers is subjected to a period of drought. A mutation that allows the population to survive this condition could quickly spread through the gene pool (complete collection of genes in a population). If, however, things changed and cool wet weather set in, the entire population might be lost. In sexually reproducing populations, the high variation in such types may help protect against such devastation.

Sex in plants

A new individual of most plants starts life as a cell made by the fusion of an egg cell and a sperm (or its equivalent). Some of the simpler water plants, the algae, discharge these sex cells into the surrounding water so that the free-swimming sperms unite with the much larger egg. The resulting product of fertilisation, then grows into a complete plant. Mosses and ferns also rely on the presence of a film of moisture to allow the sperm to swim to the egg, which is housed in flask-like structures on the plant. In these plants, fertilisation is internal but still relies on water being present. It has resulted in the distribution of these types being restricted to areas which have good water supplies, at least at some times of the year.

True flowering plants are not so dependent on water for reproduction and are therefore found in a much greater variety of habitats compared to their more primitive ancestors, the ferns and mosses.

The male sex cells of flowering plants are formed in the pollen grains. The female cells form in a container called the ovary, surrounded by petals which may be large and obvious or very small and inconspicuous as in the grasses. The pollen cells reach the ovary through the stigma, an organ located above the ovary.

The stigma is covered by a sticky material that stimulates the pollen grains to germinate if they come from the same species of plant. Upon germination, the pollen grains grow long tubes that pass down to the ovary. The male nuclei from the pollen grains pass through the pollen tubes into the ovule and unite with the female nuclei. The fertilised cells are supplied with food and tough outer coats. Then they are released as seeds. Some seeds lose their power to germinate in a few days, but most survive for three to 15 years if they are kept dry and cool. Some seeds from species such as the lotus flower have germinated after 100 to 150 years!

Co-operative partners

Pollen requires no water for transport to the female parts of plants. Hay fever sufferers will be aware of the fact that astronomical numbers of pollen grains of some species are carried on the wind and are in the air that we breathe. Parallel to this method of pollen dispersal, another technique developed. As flowering plants were evolving on Earth, so were other species. Many of these began to be dependent on one another in increasingly intricate and complex ways. As insects were spreading out, invading new areas, taking advantage of what the environment provided, some began to exploit flowers. While feeding on one plant and then another, they became carriers of pollen. As time passed, some flowers would have developed an increasing dependency on the visits of pollen-bearing insects.

In time, plants began to compete for the attention of insects through natural selection. They developed increasingly attractive flowers that insects can see, sweet nectar, and shapes that allowed the

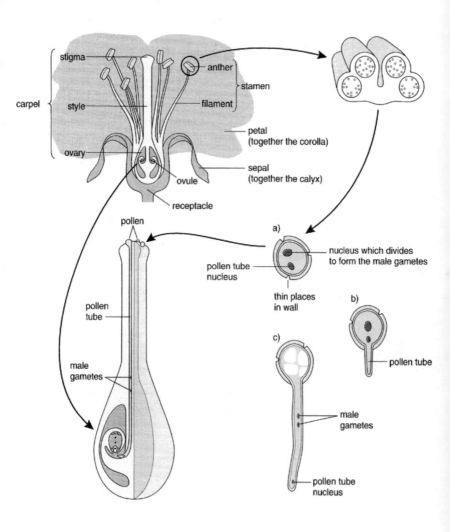

Figure 2.2   Sex in flowers

insects to land on them. Those that attracted more insects left more offspring, and some groups of flowers became true specialists at attracting certain specific insects. The insects, on the other hand, became increasingly specialised so that they could enjoy the offerings of the specific sorts of plants that yielded the nutrients most closely fitting their specific needs. The insects and the flowers changed in ways that enhanced their interdependency and specialisations in a steady progression toward finer attunement to each other. Such reciprocal influences on the development of interdependent species is called coevolution.

Sex in animals

People are often under the impression that sex is sex, that basically all species must do it the same way. Nothing could be further than the truth. Some animals do behave in ways that humans would recognise, but most of them don't. For example, bedbugs are often homosexual, however, neither male nor female bedbugs can survive many matings because the male pierces an organ in his mate's back with his sharp penis and ejaculates directly into the body cavity. Special cells then capture and ingest many of the sperm as they roam the recipient's tissues. Thus the recipient is nutritionally rewarded. Other species may also surprise us. In some mites, brothers and sisters copulate before they are born, so the little females are born pregnant and undoubtedly disillusioned!

Certain snails have an enormous penis just over the right eye. They are hermaphrodite (each possessing both male and female sex organs), but they cannot exchange sperm until they have pierced each other with a chalky dart that usually acts as a sexual stimulant but can also kill. Other snails begin life as wandering males, but eventually settle down and become sedentary females. A wandering male can only mate with a female before his masculinity fades. The female praying mantis sometimes makes a meal of a male even while he is copulating with her. In fact, after she eats his head and brain, his sexual activity becomes more intense – but we should not extrapolate from that to human behaviour!

Some snails, fish and lizards change gender, so they may be a father at one time and a mother at another. In yet other fish, there are no males at all. The egg is stimulated to develop when the female mates with a male of a different species. His sperm only activates the egg, it does not join with it.

On the more kinky side, by human standards, we find geese that form ménage à trois (a threesome, usually with two males). 'Unwilling sex', incidentally, is common in lobsters, skunks, camels, rhinos and orangutans, where females are subdued before they will allow mating.

The reproductive structures of many animals can be somewhat surprising. Males of many insects have penises that look like instruments of torture with points, hooks, barbs, and impossible angles. That of each species is so distinct that breeding between species is not possible. Male opossums have split penises with grooves instead of tubes, that match the divided vagina of the females. Pigs have corkscrew-shaped penises that lock tightly in the female's vagina. Snakes and alligators can copulate with either of two penises that evert from the cloaca, like turning a finger of a glove inside out. Each is covered with backward-directed barbs, as are the male organs of cats and skunks. Also, as all the worst comedians say, hedgehogs and porcupines have to do it very carefully!

Life after sex

Imagine a pile of sand and cement, mixed up with some clay and the other ingredients of building bricks. You pass by and throw into this mixture one complete brick. Then as you watch, a wave of activity sweeps over the pile. Other bricks form and arrange themselves into an intricate ornamental garden wall. This wall is not only self-made, but self-maintaining. If a brick loosens and falls out, the wall makes another and replaces it. This analogy is applicable to living organisms, which are of several orders of magnitude more complex than one layer of building bricks. When a human egg is fertilised, it too is swept with a wave of activity. The end product is a self-organised being of billions of individual cells formed into bone, muscle, nerves and all other tissues that we possess.

As soon as a sperm's head enters an egg, a transparent fertilisation membrane forms around the egg and acts as a barrier to other sperm cells. The fertilisation membrane eliminates the possibility of abnormal growth that would result if two sperm cells fused with the egg's nucleus. It is only the body ('head') of the tadpole-shaped sperm that penetrates the egg. The 'tail' is discarded, while the head, which is nothing at all but nucleus, moves through the egg's cytoplasm to fuse with the egg's nucleus. Fertilisation is now complete; a new cell has been formed with half its nucleus contributed by each parent.

The sperm activates the egg and initiates cleavage – a process of cell division called mitosis. The cell divides once to produce two, and then again to produce four, eight, 16 cells and so on. Occasionally, one of the early cleavages may divide the embryo into equal parts which separate. When this happens identical twins, triplets, or even quadruplets will develop independently. After several divisions, the cluster of cells looks like a mini blackberry – hence its name, *morula*, from the Latin for blackberry. It becomes a hollow ball-like structure which eventually becomes an embryo. In the present-day research programmes involving human embryos, the point at which we should call the collection of cells a human individual is conjectural and emotive. Certainly, at this point, no tissues or organs are present. The cells are undifferentiated.

The ball of cells becomes implanted in the wall of the uterus and division continues, but some cells now divide faster than others. The differential rate produces strain that causes one side of the ball to push in at one side to form a hollow two-layered cup of cells called the gastrula. The embryo is now at a critical stage. From this point on, the cells will be headed toward their future state. Some are to become bone, others, muscle, some will be nerve cells.

The process of differentiation begins with the layer of cells forming inside the cup-shaped gastrula. The inside of the cup, the endoderm, begins to show characteristics quite different from the outside, the ectoderm. Soon, a third layer, the mesoderm, grows between the two.

The extraordinary fact is that in all vertebrates, the same kind of tissues develop from the same 'germ' layers. A bird's feathers, a snake's scales, the slimy skin of a frog, and the soft skin of a human baby all arise from the ectoderm. The question is, how does this sculpturing of living material occur?

From hundreds of investigations, scientists hypothesised the existence of hormone-like chemicals called organisers. According

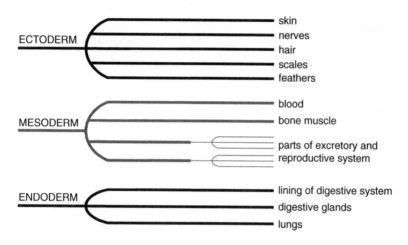

Figure 2.3   Primary germ layer from which all tissues develop

to the organiser theory, these chemicals can affect genes, which in turn affect the rate of development of tissues and organs. They switch genes on or off at certain stages of growth. More distressingly, some chemicals can interfere with organiser chemicals, either directly or indirectly, by affecting the controlling genes. Thalidamide is a tragic example of this and manifested itself in the 1960s when it was prescribed to pregnant women without knowledge of its consequences.

The garden of inheritance

Acorns always grow into oak trees and never into elms. Dogs always have pups and never lambs. We all expect that every living thing will have offspring just like itself. But why? And why also do black sheep sometimes have white lambs; dark-haired, brown-eyed parents sometimes have blond, blue-eyed children? We know that as cells multiply, each new cell is connected to the cells that preceded it by substances in the nucleus. It seems reasonable that the answer to our questions must lie in the nucleus – and so it does. The physical characteristics of any organism are determined by the following factors:

- **Heredity** — each individual inherits from its parents its basic structure, species and gender
- **Variation** — the plan transmitted from parent to offspring is rarely exact. The plan for each individual has an extremely high probability of being just a little different from its parents' plan, and as such, being unique
- **Environment** — an organism's surroundings supply materials and influences that bring the inherited plan to full development

Genetics is the branch of science that studies the similarities and differences between parents and offspring. For hundreds of years, farmers knew that if they bred from cows that gave the most milk, or wheat with the largest grains, they were likely to get these desirable characteristics again.

The results of such breeding, however, were not guaranteed. They often did not turn out as expected. Sometimes the offspring had the desired characteristics; other times they did not. Desired characteristics which did not appear in the offspring sometimes reappeared in the grandchildren. It was all very mysterious, and questions about heredity continued to be shrouded in mystery and folklore until the 1860s, when an Austrian monk, Gregor Johann Mendel (1822–1884), provided the first scientific clues to the way that traits are inherited.

Mendel formulated two laws to explain the pattern of inheritance he observed in crosses involving the common garden pea, *Pisum sativum*. The first law, the Law of Segregation, states that any character exists as two factors, both of which are found in the body cells (somatic cells) but only one of which is passed on to any one gamete. The second law, the Law of Independent Assortment (see p.93), states that the distribution of such factors to the gametes is random. Therefore, if a number of pairs of factors is considered, each pair segregates independently.

Mendel gave the name 'germinal units' to those features that he considered to control characteristics in his experimental pea plants. Today these germinal factors are called genes (a term first used by the Danish Wilhelm Johannsen in 1909). The different forms of genes are called alleles, an abbreviation of 'allelomorphic pairs of genes'. It is known that a cell with the full complement of chromosomes for that species (diploid cell), contains two alleles of any particular gene. Each allele is located on one of a pair of homologous chromosomes (chromosomes that pair during meiosis). Only one homologue of each pair is passed on to a gamete (an egg or a sperm). Thus, the Law of Segregation still holds true. Mendel envisaged his factors as discrete particles but it is now known that they are linked together (see p.97) on chromosomes. The Law of Independent Assortment therefore only applies to pairs of alleles found on different chromosomes.

Gregor Mendel

Gregor Mendel was born in 1822 and christened Johann Mendel. His family was of peasant ancestry and lived in Moravia, now a province of the Czech Republic. After simple, early instruction by an uncle, Mendel attracted the attention of the clergy and, at the age of twenty-one, became a monk in the monastery of Brunn. At his ordination, he took the name of Gregor, by which he has been known ever since.

Mendel attended the University of Vienna from 1851 to 1853, where he studied mathematics and physics, and then became a teacher of science at the local secondary school. He continued to teach until 1868, when he was made abbot of the monastery, a position of such responsibility that he did no further teaching and gave up his research as well. It is interesting to speculate on what might have happened had the scientific world responded to Mendel's two publications which led ultimately to the foundation of the science of genetics. Mendel was discouraged by the almost complete indifference of the scientific community to his work. It must, indeed, have been a prime factor in his failure to continue his experiments.

For a period of about eight years, from 1858 to 1866, Mendel conducted his famous breeding experiments with the common garden pea for his long and meticulous studies. His first paper was published in the *Transactions of the Natural History Society of Brunn*, the obscurity of which may account for the indifference with which the paper was received. Thirty-four years later, in 1900, three scientists from different countries independently discovered Mendel's work and realised its significance. They were H. De Vries from Holland, C. Correns from Germany, and E. Tschermak from Austria. They completely confirmed Mendel's findings which were to become the foundation stone of the whole of genetics which has become such a dynamic science in the twenty-first century. Mendel died in 1884 before his work was recognised for what it was. Today, we call him 'the founder of heredity' or even 'the founder of molecular biology' and the first mathematical biologist.

Mendel's first law

Mendel's first law was formulated as a result of his crossing plants with a single pair of contrasted characters. He used this method to determine his Law of Segregation. He crossed parental pea plants which were pure bred for producing round seeds, with pea plants that were pure bred to produce wrinkled seeds. He found that all the hybrids in the F1 produced round seeds, thus showing the dominant character, round and not the recessive, wrinkled. He allowed these to self-pollinate at random, and found that the missing recessive wrinkled trait reappeared in some of the F2 plants. Moreover, the ratio of the F2 plants with the recessive trait to those with the dominant trait was fairly constant, regardless of which contrasted characteristics were involved. The ratio was 3 dominant: 1 recessive, strangely enough, a detailed statistical analysis of Mendel's results made by Ronald Fisher in the 1930s showed that Mendel's figures from his later investigations were, in statistical terms, too good to be true! The implication is that once Mendel had formulated his theory, those results that failed to give the expected ratio were ignored by Mendel or by the gardeners who helped him count his seeds or plants.

The details of Mendel's cross involving one pair of contrasted characters are given in the diagram opposite:

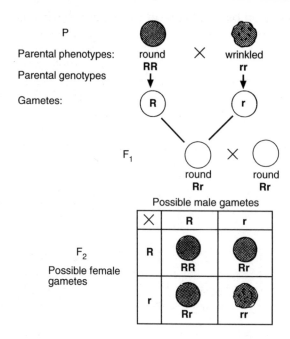

Figure 2.4   A monohybrid cross using a Punnet square

Mendel's second law

In 1859, Gregor Mendel (see p.91) carried out experiments involving two pairs of contrasted characters to check whether traits were inherited independently of each other. He used pea plants that always gave round, yellow seeds and crossed them with others that always yielded wrinkled, green seeds. If these four traits could be shuffled around quite independently of each other, he would get every possible combination in the first hybrid generation. If these F1 plants were self-pollinated, he obtained a ratio of 9 round yellow : 3 round green : 3 wrinkled yellow : 1 wrinkled green.

His results are illustrated below and form the basis of his law of independent assortment.

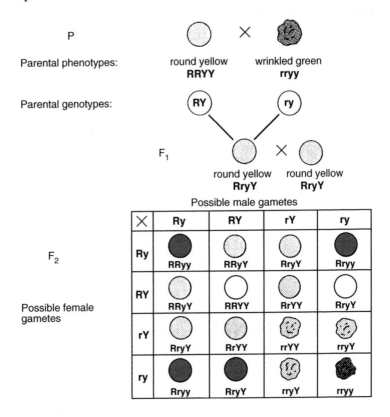

Figure 2.5   A dihybrid cross using a Punnet square

Sex cells and fertilisation

The largest known cells are eggs. They contain little but cytoplasm; their bulk is mostly water and stored food. The important part of an egg is its nucleus. Sperm cells are hundreds or even thousands of times smaller than eggs. They are little else but a nucleus attached to a vigorous, whipping tail, the flagellum. When the sperm finds its goal, its nucleus joins the egg nucleus to form the nucleus of a new and separate cell.

The nuclei of both kinds of sex cells generally have fewer chromosomes than the other cells in the body of the same organism. The reason for this becomes clear when we see what would happen if they did not. Human body cells have 46 chromosomes. If each matured sex cell, or gamete, also had 46, then a baby would have 92, and its children would have 184. Yet all normal cells in the human body have 23 pairs – 23 from mother and 23 from father. Cell nuclear division by mitosis ensures that each new cell gets a full set of chromosome pairs.

Microscopic studies of developing human sperm cells and eggs show that they have only 23 chromosomes, one from each pair. How does an organism produce cells with half the number of chromosome pairs?

Of all the billions of cells in our bodies, and those of other sexually reproducing organisms, only cells in the reproductive organs can divide so that they split up the chromosome pairs. Logically enough, the process is called reduction division, or meiosis. To see how it works we can follow meiosis as it occurs in an organism which has only two pairs of chromosomes. (see p.97).

Figure 2.6 shows meiosis as a series of stages but, in reality, one stage merges into the next in a continuous process as elaborate chromosome manoeuvres take place. In the early stages the chromosome pairs move towards each other (B) and appear united along their whole lengths (C). At this point the chromosomes are so close together that the nucleus looks as though it has only half its normal set.

The chromosomes now shorten and thicken and divide lengthwise into two, to form twin threads. The double pairs now begin to separate (D), though the separation is not complete because at one or more points along their lengths the threads of the double chromosomes remain in contact. Odd shapes are produced as the chromosomes cross over each other. These points of contact are called chiasmata (from the Greek '*chi*' or cross). At this point, the half chromosomes can break at the points of cross over and rejoin in different ways, thus exchanging genes that have come from the

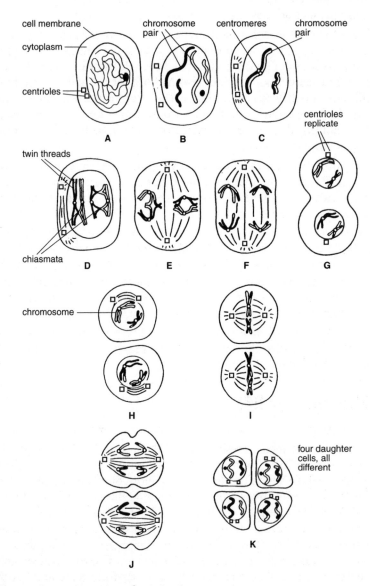

Figure 2.6 Stages of meiosis

father with those that have come from the mother. The result will be *new* chromosomes, i.e. chromosomes with different genes from those present originally.

Next the double chromosomes separate from each other (E) and move to opposite poles of the cell (F). A nucleus then forms at each pole, and the cell divides into two (G). At this point there are two daughter cells each containing chromosomes made up of twin threads. Finally these double chromosomes separate and each daughter cell divides into two, this is mitosis (H–K). This results in four cells in all, each with half the normal number of chromosomes.

When the male sperm unites with the female egg in fertilisation, each provides half the chromosomes for the new individual. Since it is highly unlikely that there ever existed a cell with this identical set of chromosomes, each fertilised egg is unique. The advantage of sexual reproduction lies in the uniqueness of the fertilised egg that makes each new life just that little different from either of its parents. Variation among offspring may produce one or more that may be able to adapt to changing conditions of the environment.

Once an egg has been fertilised, all subsequent cell divisions are mitotic and each new cell receives the full number of chromosomes. Eventually, the new organism will reach maturity and its time will come to make a contribution to the survival of its kind. Its reproductive organs will then produce sperms or eggs and the cycle of life will have made a full turn.

Threads of life

With our present-day knowledge of chromosomes and meiosis we can explain Mendel's results, using a capital letter for a dominant gene and a lower case letter of the same kind for a recessive gene. Thus T stands for the tall factor in pleas and t represents the short factor. Although garden peas have seven pairs of chromosomes, we will work only with the pair that carries the gene for height.

Think of two peas in flower, ready to produce reproductive cells. One is a tall plant. Both of its chromosomes carry the tall gene, T. The chromosomes of the other plant, a short one, each carry the short gene, t. It is now time for the cells to form male and female gametes. Meiosis occurs and the chromosome pairs are divided by reduction division. The results are male cells, each with a chromosome bearing the T gene; and female cells, each with a chromosome carrying the short gene, t.

When the cells join to form a fertilised egg, the reconstituted chromosome pair will have one T chromosome and one t chromosome. It will appear tall, because T dominates t. We can also see that if the original tall plant had been crossed with a tall plant carrying only tall genes, all the offspring would have TT chromosomes and would also grow into tall plants. A cross between the short plants would have similar results. That is why plants that always breed true to type are called pure-bred.

But let's go back to the plant Tt, with its mixed genes. When its cells go through meiosis it can form two kinds of reproductive cells, one with a chromosome carrying a tall gene, the other with a chromosome carrying a short gene. When two plants of this type are crossed, we can get three kinds of combinations of genes in the fertilised eggs: TT, Tt, and tt. This explains Mendel's law of segregation.

During his eight years of experiments, Mendel worked with seven pairs of contrasted characters and found that they all combined and recombined in exactly the same way. Since it is now known that peas have seven pairs of chromosomes, we can see that each of the characteristics Mendel studied was located on a different chromosome pair.

That seems quite logical until we recall that humans have 23 pairs of chromosomes and fruit flies have four pairs. Geneticists have studied dozens of fruit fly characteristics, and everyone knows that humans have more than 23 characteristics. There is only one way to explain this and it is to assume that each chromosome has many genes linked to it. By the year 2000 the human genome project enabled scientists to make a map of all human genes.

The human genome project

Since the 1980s, advances in the fields of both genetics and medicine have led to developments in medical genetics which have allowed medical practice to evolve at a rapid pace. The human genome project is an international research effort that has succeeded in analysing the structure of human DNA. Scientists have mapped the location of an estimated 100,000 genes. It is anticipated that the end product of this research will be the standard reference for biomedical science in the twenty-first century and will help us to understand, and eventually treat, many of the four thousand plus recognised human genetic afflictions. The aims are:

- mapping and sequencing the human genome (all human genes)
- mapping and sequencing the genomes of certain other organisms
- data collection and distribution
- research training
- international sharing of ideas in gene technology

Initial estimates suggested that all this would take up to 15 years. In fact the completion of the initial map took place early in 2000, which was before the predicted time schedule. The purpose of the enormous research programme was to sequence the four bases making up human DNA, i.e. adenine, cytosine, guanine, and thymine in human DNA. (see DNA structure p.102). The bases are represented no less than three thousand million times in our genome.

If typed in order, using their initial letters, A, C, G, and T, our sequence of bases would fill the equivalent of 134 complete sets of the *Encyclopaedia Britannica*. The size of an individual gene within the whole length of human DNA is similar in comparison to the size of an ant on Mount Everest! This mind-boggling project will provide an invaluable reference for medical science in the study of human genetical disorders throughout the twenty-first century.

More light on chromosomes

Some of our initial questions about heredity have been answered. We have a sound explanation of why offspring resemble their parents and why white sheep sometimes produce black lambs. Why, however, are some of us males and others females?

In the 23 pairs of chromosomes, there are 22 matching sets and an odd pair called sex chromosomes, X and Y. Actually they look different; the human Y chromosome is shorter than the X chromosome. The Y chromosome in fruit flies has a little hook, while the X chromosome is quite straight in comparison. It has been found that cells from male mammals, including ourselves, have one X and one Y. Female cells have two X chromosomes. After reduction division, egg cells can contain only one X chromosome each, but sperms can contain either an X or a Y. If an egg is fertilised by a Y-bearing sperm, the offspring will be male. When two X-bearing reproductive cells unite, the offspring will be female.

The genes that are easiest to locate in humans are those linked to the X chromosome; these are sex linked genes. One of these controls our ability to see certain colours. The gene that produces colour blindness is recessive to the gene for normal vision. Thus a colour-blind woman must have the colour-blind gene on both X chromosomes. Men can be colour-blind with only one abnormal gene, since the Y chromosome is incapable of carrying either a gene for normal vision or for colour-blindness.

It is now clear that genes are inherited independently of each other only when they are located on different chromosomes. If genes are linked together on the same chromosome they are inherited together. Occasionally, however, scientists find an individual in which genes that were apparently linked in the organism's ancestors seem to have separated and relocated on some other chromosome.

Occasionally something goes wrong during cell division; a cell with an abnormal number of chromosomes is produced. If the number of chromosomes is an exact multiple of the normal number, such cells form a polyploid organism. Garden plants with double flowers are polyploids.

Sometimes, in humans, an abnormal chromosome number can cause problems. Failure of chromosome pairs to separate at meiosis results in individuals with physical and, or mental abnormalities. Down's Syndrome is an example of this, when chromosome pair number 21 fails to separate and the afflicted individual has 47 instead of 46 chromosomes. In some individuals there is only one X chromosome. This is Turner's Syndrome and results in individuals who are under-developed females. The chromosome pattern XXY produces under-developed and mentally retarded males and is called Klinefelter's Syndrome.

Life's tool kit

Life is a series of chemical processes and of all the chemicals found in living things, the complex molecules we call proteins are the most distinctive and essential. It is not surprising then that quite early in the history of genetics, scientists knew that genes must be chemicals and guessed that they were probably proteins. What better place to begin the search than the nucleoproteins that were identified as the substances of which chromosomes are composed?

By 1950, however, it became quite clear that it was not the protein component of chromosomes that had the important role of passing the secret code of life from generation to generation, but another component, nucleic acids. These are the largest and by far the most fascinating of all life's molecules. Two forms of nucleic acid are known: deoxyribonucleic acid, DNA, is found in all chromosomes of all cells of all living organisms; ribonucleic acid, RNA, is found the cytoplasm of cells and in the nuclei of many.

Nucleic acids, like proteins, are made of many units strung together. They have a 'backbone' chain of sugar and phosphate molecules, to which is attached another compound, a nitrogen-containing base. DNA has two of these spirals wound around each other. The two chains are held together by the attraction of hydrogen bonds.

Proof that the DNA of chromosomes is the transmitter of genetic information has come from many investigations. Biochemists have measured the amount of DNA in cells and found that every body cell nucleus of the same species has the same amount of DNA. Sex cells have half the number of chromosomes and exactly half as much DNA.

The pioneering work which led to the proof that DNA carries genetic information between cells started with a British public health officer, Frederick Griffith in 1928. He was studying pneumonia, a widespread disease responsible for many deaths at the time. Pneumococci, the bacteria that cause some forms of pneumonia, come in a variety of forms. There is one type that exists

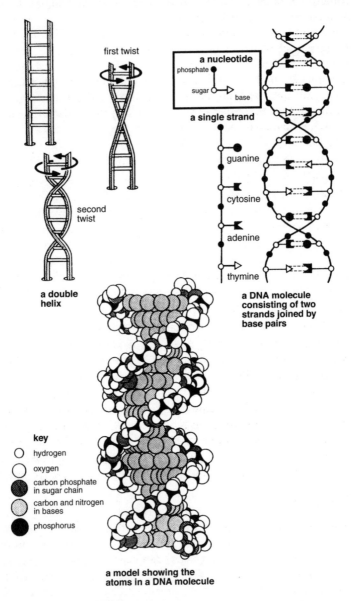

first twist

a nucleotide

phosphate

sugar ●——▷ base

a single strand

guanine

cytosine

adenine

thymine

second twist

a double helix

a DNA molecule consisting of two strands joined by base pairs

key

○ hydrogen

○ oxygen

● carbon phosphate in sugar chain

○ carbon and nitrogen in bases

● phosphorus

a model showing the atoms in a DNA molecule

Figure 2.7   DNA structure

sometimes as cells enclosed in a smooth capsule and at other times as a rough, non-capsulated strain. The smooth strain (S) causes pneumonia, the rough strain (R) does not. If the R-strain of bacterium is grown in a mixture of materials containing DNA from dead S-strain bacteria, some cells of the harmless R bacteria are transformed into the dangerous S type. When they reproduce, these bacteria continue to produce disease-causing bacteria with capsules. This transformation of one type of bacteria to another can be explained only by assuming that the change was caused by DNA, the carrier of inheritable traits.

Of all the forms of life – if they are indeed a form of life at all – none have been so puzzling or so important to DNA research as the viruses. These particles are far smaller than bacteria and can be seen only with the electron microscope. However, before the invention of this instrument, Wendell Stanley, an American biochemist, found that a virus consists only of DNA or RNA surrounded by a coat of protein. It makes a living by injecting its DNA into cells and over-powers them by altering the chemical processes that they normally carry out so that they make more virus DNA. The virus DNA then uses the amino acids in the cell to produce new protein coats for itself. They are, therefore, the ultimate parasites.

The study of viruses and of the way they usurp control in cells, reveals a great deal about nucleic acids. It tells us that there are different kinds of DNA, each capable of producing a typical kind of virus, with its own characteristic shape and protein coat. It also tells us definitely that DNA can make more molecules like itself. Self-duplication is, of course, an absolute requirement of the hereditary molecule, otherwise the reproductive cells would have to contain all the DNA found in the adult organism. In fact, reproductive cells have only half the amount of DNA found in one single body cell.

How genes work

The discovery that DNA carries the code of life was a gigantic step forward. The next step was to discover how the molecule duplicates itself, but that required knowledge of its structure. In the early 1950s scientists all over the world were attracted to the problem of DNA structure.

The most probable model of the DNA molecule was made by James D. Watson, an American, and Francis H. C. Crick, an English scientist, who worked together at Cambridge University. They proposed the double coil, or double helix structure. (see Fig. 2.7).

When the structure of DNA became clear, it was possible to imagine how the molecule might reproduce itself. The bases that join the two chains (like rungs across a ladder) are of four kinds; guanine, cytosine, thymine and adenine. Guanine always links with cytosine, and thymine always pairs with adenine. The differences between the DNA of all organisms, between and within species, depend completely upon the pattern of rungs.

When chromosomes are undergoing mitosis, the DNA molecule is 'unzipped' using an enzyme. Each half is then free to serve as a pattern on which new bases, available in the cell, will attach themselves.

Because adenine joins only with thymine, and guanine will pair only with cytosine, each half builds the duplicate of its lost partner. Thus, at the end, there are two DNA molecules which are exact duplicates. But how does DNA build proteins? Actually, it doesn't. The function of DNA is to form RNA (ribonucleic acid), which differs from DNA by having as one of its bases, uracil instead of thymine. After unzipping, each helix of DNA acts as a mould for matching RNA and this acts as a messenger to the cytoplasm. The messenger RNA leaves the nucleus and attaches itself to small bodies known as ribosomes (see p.10) in the cytoplasm. The ribosomes serve as assembly points for amino acids to be pieced together according to the messenger RNA code.

Proteins are built of only twenty different amino acids. The differences between proteins depend on two things: the number of amino acids and their sequence within the protein.

We can imagine that the sequence of bases in the RNA molecule makes up a set of coded instructions that in effect say, 'take three molecules of the amino acid, glycine and add two molecules of phenylalanine (another amino acid), repeat ten times and then add three molecules of the amino acid, alanine and so on ...'. Thus DNA makes RNA and RNA makes protein.

Proteins are the structural materials of cells, but the enzymes that control the chemical reactions of cells are also proteins. It is now clear that DNA not only controls the building of cells, but it directs their activities as well.

The alphabet of life

There is a genetic alphabet which is used to translate the linear sequence of bases in DNA (see p.102) into the sequence of amino acids in proteins. The alphabet is A for adenine, C for cytosine, G for guanine and in RNA, U for uracil. An alphabet of only four bases is enough to code for the thousands of different proteins that exist in living organisms. Although proteins are complex molecules, they are built from only twenty different amino acids (see p.26).There is a simple mathematical explanation. If each base codes for one amino acid, only four amino acids can be specified. A two-base combination would provide four times as many possible combinations (4 x 4 = 16), but this is still not enough. By 1952, which was quite early in the search for the genetic code, the biochemist, Alexander Latham Dounce, predicted a three-base combination to code for each amino acid. The prediction was confirmed by George Gamow, a physicist, two years later. It was reasoned that a combination of any three of the four bases would provide 64 possible triplets (4 x 4 x 4 = 64). However, there are only twenty amino acids, so more than one triplet of bases could code for the same amino acid. Also, some code as 'full stops' to terminate a chain of amino acids when a protein has been made, and others can act as a 'capital letter' to begin another sequence of amino acids. Each triplet of bases is called a codon.

Gene technology

The science of genetics is essentially a product of the period since the 1880s, but the rate of scientific advance in gene technology during the last decade of the twentieth century has been truly remarkable. Largely because of a lucrative partnership between geneticists and industrialists with entrepreneurial skills, genetic engineering is undergoing a revolution.

The term, genetic engineering is familiar to most people today, having been used in both emotive and trivial contexts by the media and also by writers of science fiction. Indeed, science fiction and some purported 'serious' accounts of experiments in genetic modification by some newspapers are difficult to tell apart. Gene technology subsumes many concepts, including gene manipulation, gene cloning, recombinant DNA, gene therapy and genetic engineering. Very briefly, the earliest type of genetic engineering relied on finding specific genes, cutting them out of chromosomes

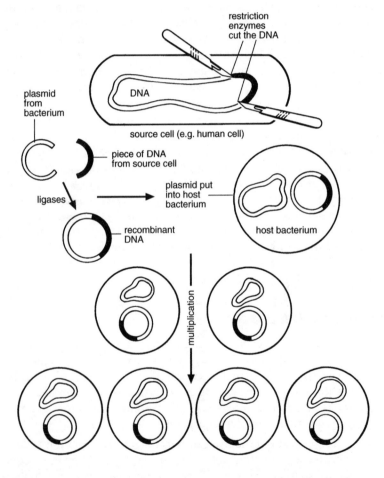

Figure 2.8   The principle of genetic engineering

and splicing them into chromosomes of microbes like bacteria or yeast. After the genes had replicated many times, the proteins made by them could be harvested.

The main advantage of this technique is the mass-production of many otherwise scarce and expensive protein-based chemicals which have a direct role in saving lives. For example, the extraction of just 5mg of the protein, somatotrophin (a growth regulating hormone) would require a half a million sheeps' brains! The same mass of this hormone can be made in one week by nine litres of genetically engineered bacteria.

The genetic engineer uses some highly specialised tools and terminology:

- the source DNA contains the required gene, which is cut out and added to the host DNA
- the host DNA is cut with enzymes to allow the insertion of the source DNA
- the recombinant DNA is hybrid DNA resulting from the fusion of the source DNA and the host DNA

The potential for mis-use

So the mass-production of life-saving proteins through genetic engineering is already with us, but what about the potential life-threatening products of gene manipulation? For example, what would happen if someone inserted a cancer-causing gene into the familiar bacterium, *Eschericia coli*, which lives harmlessly in our intestines, or a gene that makes a deadly toxin? What if these bacteria were then released into the human population? This may seem unlikely, but history books now tell us about mustard gas of World War I, the atom bombs of World War II, napalm in the Vietnam war and nerve gases of other recent wars – someone invented them and someone used them! Biological warfare is not well documented in popular science publications because most developments are classified as top secret, but behind the closed doors of secrecy, anything is possible.

Another less cynical concern about genetic engineering is that well-meaning scientists could accidentally allow a potentially dangerous genetically engineered organism to escape into the environment. Even after smallpox was supposedly eradicated from Earth, there were two minor epidemics in Europe caused by cultured experimental smallpox viruses that had escaped from a laboratory. The result was that one person died of a disease that technically did not exist!

Genetic engineers have always been aware of this accident waiting to happen and have safeguarded against it. In the same way that they can produce recombinant DNA which carries genes for manufacturing useful products, they can introduce genes into that same DNA which make it possible for the microbes to survive only in highly specific conditions. In other words, if these altered microbes were to escape, they would certainly die. In the early days of gene manipulation using *E. coli*, a mutant was produced that had a defective gene preventing it from making its usual protective cell coating. The material that the gene normally made had to be provided artificially in the culture medium. However, microbes mutate naturally on their own. There is a biological rule which states that, given a sufficiently large population and sufficient time, organisms will adapt to selective pressures of the environment in such a way that it is to their advantage. In this instance, the microbes soon mutated back to the form that could make their own coats. Another gene responsible for the production of the coat had to be deleted. Even this was not enough; they reproduced anyway.

The problem was investigated by two American scientists, Roy Curtiss III and Dennis Pereira. The latter found that the microbes were manufacturing a sticky substance called colanic acid which acted as a protective coat. When they destroyed their ability to produce this chemical by mutating the offending gene, they finally had a tailor-made bacterium which could not exist outside a highly specialised artificial environment. Unexpected bonuses are often products of research and this was no exception – the new microbe was also sensitive to ultraviolet light, so would be killed by sunlight.

One problem remained. About one in a million bacteria is sexually active and reproduces by mating (conjugating) with another. So, theoretically, if an escapee mutant *E. coli* mated with a normal bacterium, any dangerous gene could be passed on to found a colony of unwanted mutants. Curtiss solved the problem by altering yet another gene in the mutant, one which normally controls the manufacture of thymine, an essential constituent of DNA. Thymine therefore must be supplied in the cuture medium. The microbe was finally crippled and rendered totally dependent on laboratory conditions.

Threats or opportunities?

It is interesting to look back to the 1970s when attitudes to genetic engineering were quite different from those of today. In February 1975, scientists met in Asilomar, California, and held a heated discussion to draw up guidelines which would prevent the nightmare situation of genetically engineered mutants escaping. Rules were formulated and some nations banned research on recombinant DNA techniques altogether. Later in the 1980s, attitudes changed and much of the opposition to research and development in this area dissolved. Was it a coincidence that this acceptance occurred in parallel with the realisation of massive commercial feasibility and profit in this field?

There are people who still fear the results of gene manipulation but no one can deny the promise of the technique to mass-produce life-saving chemicals in the twenty-first century and beyond. This line of research started in small laboratories, often staffed by post-graduate students focussed on a brave new future. Multinational drug companies then reacted with great enthusiasm, making stock brokers millionaires over night and tempting many a dedicated scientist to leave the ivory towers of academia.

Many possibilities emerged for recombinant DNA technology, not the least of which was the potential for mass-production of hormones. Insulin, for example, could be obtained only from animal tissues and in relatively small amounts in the past. Today, almost unlimited quantities of insulin can be harvested from genetically engineered microbes, protecting the lives of many diabetics throughout the world. Immunity to certain viral diseases can be enhanced with interferon and interleukin which can now be produced in large amounts via genetic engineering.

Labelled for life

The fact that the sequence of nucleotides in our DNA (see p.102) is as individual as our fingerprints, is the basis of a new method of identification which can be used in forensic science. This, so-called genetic fingerprinting was first used as evidence in the late 1980s to establish the guilt of a murderer. Since then the technology has been the subject of intense scrutiny in courts of law and in the media. The term, genetic fingerprinting was soon coined to describe its use in criminology, but patterns on the skin of the fingers have nothing to do with it! The analogy is the uniqueness of both a set of fingerprints and a DNA profile. The technique is particularly good for identifying culprits of violent crimes where body fluids such as blood or semen are left at the scene of the crime. There is now a computerised database holding records of all convicted criminals and this promises to be the greatest breakthrough in criminal justice since the discovery that fingerprints are unique to individuals. Forensic science laboratories have allocated significant resources to its further development.

Of course, the use of genetic 'labels' is not new. The first attempts were via blood groups after the A B O system of blood classification was discovered by Karl Landsteiner of Vienna in 1900. However, to successfully identify blood groups, there had to be a sufficient quantity of blood that was not degraded. Crimes were often frustratingly left unsolved because of the uncertainty of the analysis of a poor sample of material.

A breakthrough came in the mid-1980s when Professor Alec Jeffreys at Leicester University showed the presence of many variable regions of DNA which did not code for amino acids. These regions were called minisatellites and there are thousands of them scattered throughout the chromosomes, probably having evolved as mistakes during replication of DNA (see p.104). If enough regions of DNA are examined, it is possible to obtain a genetic profile which is almost unique to an individual. The relatively simple technique of analysis provides an X-ray film with a pattern of bands showing the relative positions of minisatellites, rather like a bar code on shop-bought materials.

In the first test case using genetic fingerprinting it was shown that a suspect who had confessed to a rape and murder was in fact innocent. Two girls had been raped and murdered in the same part of the country, but the crimes were committed three years apart. Investigators suspected a connection between the two and soon a suspect was cross-examined. He admitted to one of the crimes but denied being involved in the other. Forensic scientists used DNA profiling which showed clearly that both crimes were committed by the same person, but the suspect was not that person. All men from the area were screened and genetic profiles made of 5000 samples. Finally one sample matched those from both crime scenes and the culprit was arrested and convicted.

In the 1990s a more sophisticated technique for DNA profiling was developed. The new technology required only very small traces of DNA and the chemical reactions in the analysis were speeded up. This reduced the time taken for results to be obtained. The method simulates the replication of chromosomes, producing millions of copies of DNA in a test tube. The technique detects much smaller minisatellites than those involved in earlier attempts, and the pieces of DNA that are required are almost 100 times smaller than those needed previously.

Nothing lasts!

A problem with DNA is that, in time, it deteriorates when left at room temperature, making analysis of degraded samples unreliable. Bacterial enzymes break down DNA in the same way as they cause the decay of the rest of a dead cell. This makes the *Jurassic Park* idea just a fantasy, although it is possible to use DNA from organisms that have died several thousands of years ago, if the remains have been preserved to minimise deterioration, for example mummification or deeply frozen.

One case from the 1990s used the DNA from a 5000-year old body that was found in the Alps near the Austrian-Italian border. The body was called *Otzi* the ice-man and was discovered when a particularly warm spell for the area thawed the ice from a glacier that had been there for thousands of years. The body was found, together with ancient artifacts, but sceptics thought it was a hoax. They suggested that the body was a mummy from South America and had been put there to deliberately fool the finder. However, DNA from the body was still well enough preserved for a genetic profile to be made. The results showed that the profile was far closer to those of Northern European races than to South Americans. The body was genuine and started a political wrangle about its ownership which prevented further detailed studies of the body until September 2000. Poor old *Otzi* had his parentage questioned by many scientists. Did he belong to Austria or to Italy? Does it really matter? He has provided anthropologists with some fascinating evidence about ancient cultures.

Another remarkable study resulted in the identification of the remains of Tsar Nicholas II and his family using DNA extracted from their skeletons. They were all reported to have been murdered after the Russian Revolution in 1917 and buried in an unmarked grave. When their suspected grave was eventually discovered, analysis showed that the skeletal remains belonged to a family consisting of a mother, father and three siblings. Prince Philip, Duke of Edinburgh, volunteered to give a blood sample to prove whether the mother shared the same maternal line with him, and the results supported the view that she did. Subsequent extension of the investigation used DNA from the exhumed remains of the Grand Duke of Russia, the brother of Tsar Nicholas II. The results showed that the tsar's remains had been correctly identified.

Interestingly, the scientists were also able to show that a woman living in the USA who claimed to be Anastasia, daughter of the Tsar, was definitely not!

More recently, within the last two decades, paternity testing has been revolutionised because of the development of DNA analysis, and the technique can also help in the identification of individuals after mass disasters. Similar analyses were used for victims of the Waco siege in Texas.

All of the examples quoted so far have involved human DNA. The technique can also be applied to species other than humans. Evolutionary links between groups of similar organisms have been traced by comparing their DNA. Closely related species of small invertebrates are often very difficult to tell apart by external features, and DNA profiling has been an invaluable tool to biologists when identifying new species. The method can also be used to study the extent of inbreeding within populations and consequently make better estimates of the viability of threatened species. The trade in captive animals and plants is lucrative, and the unscrupulous have been known to illegally import endangered species although they try to convince the authorities that they have bred them in captivity. Genetic profiling has led to many prosecutions in such cases particularly in trading rare birds of prey.

Hidden clues

The diagnosis of genetic disorders in unborn children has become a very sensitive issue in modern society. As a result of a technique known as gene tracking, together with chorionic villus sampling, it is now possible to find how potentially dangerous genes are inherited. Chorionic villus sampling involves taking a small sample of cells from membranes surrounding the foetus in the womb, and enables DNA from foetal cells to be extracted and tested. Armed with this knowledge, prospective parents may choose to continue with an affected pregnancy or terminate it.

In gene tracking, the DNA profile from a particular chromosome of a person who is afflicted with a disorder is compared with the DNA profiles of the parents. If there are sufficient samples of DNA from a family pedigree, partners can be given information relating to the chances of a child being affected by a harmful gene before the child is conceived. The process used for the diagnosis relies on the same technique that is used for genetic fingerprinting.

The results of genetic tracking or screening, pinpoint a potential problem to which gene therapy can sometimes offer a solution. In this technique, medical geneticists aim to alter the genes responsible for certain inherited disorders and insert perfect genes into chromosomes to replace the imperfect ones. The new genes enable the affected cells to function properly and therefore remedy the disorder. However, this is easier said than done because of the technical problem of getting the new genes into the specific cells that need them. There are three main methods for introducing the new gene into the chromosomes of recipient cells.

The first method uses viruses to carry genes into cells. One of the first disorders to be treated in this way was adenosine deaminase (ADA) deficiency. ADA is an enzyme vital for making white blood cells in the bone marrow. To treat the deficiency, a sample of bone marrow is taken from the person and the stem cells which normally develop into white blood cells are separated out.

Copies of the ADA-producing gene are taken from a human cell and placed inside a virus which has been modified so that it cannot replicate. The modified viruses are used to infect the stem cells so that the ADA-producing gene passes into them. Having received the new working genes, the stem cells are returned to the patient where they divide and produce a continuous supply of ADA.

Another method of gene therapy is used to treat cystic fibrosis. A new gene is introduced to the secretory cells of the lungs by inhaling an aerosol spray containing perfect genes. The hope is that the genes will pass into the lung cells and switch on the production of the protein needed for the supply of normal mucus, restoring the functioning of the lungs.

A third approach consists of injecting new genes so that they can reach all of the cells of the body. The control units of the genes (gene switches) are programmed to activate the genes only when they are in the target cells. One possible use of this technique is in the treatment of skin cancer (melanoma). Melanoma cells contain a particular section of DNA called a promoter. The promoter regulates the manufacture of an enzyme that controls the production of a brown pigment, melanin. Geneticists can now produce genes that are switched on only by the promoter in melanoma cells – in all other cells these genes remain dormant. It might therefore be possible to inject someone with genes that cause cells to self-destruct, in the knowledge that these killer genes would be activated only in the cancer cells.

Genetically modified (GM) products

There must be very few people who have not come in contact with genetically modified food in some form. Public displeasure at being kept in the dark with regard to developments in this field has been expressed in many forms. Under such pressure, governments have provided us with more information in the form of labels on food to indicate if it has been made from genetically altered plants. Given these labels, the public is allowed the choice to buy or not to buy. Balanced opinion on the use of genetically modified crops must take into account the fact that the yield of food per unit area of land has to be very high to feed the world's expanding population. It has to be maintained at a high level by a combination of intensive farming, selective breeding and gene technology. As populations in developing nations grow, on a world scale, any serious attempt at feeding them cannot rely on methods that worked when the number of people in the world was but a small fraction of today's inhabitants. Genetic modification is really a speeded up and less hit-or-miss version of selective breeding which has gone on since time immemorial. Many consider gene manipulation to be taboo because it appears to be less natural than interbreeding to produce a new mutant. Little knowledge produces the fear of the unknown and again, the media takes advantage of this to sell their products with emotive headlines. However, the end product of genetic modification is just a mutant – it just takes a shorter time and is a surer method of producing the desired result.

Since recorded history, humans have been cultivating plants for food by selective breeding and the same is true of the domestication of animals. Selective breeding aims to produce plants and animals progressively better suited to human needs. Until relatively recently, this form of breeding by artificial selection was unreliable. Before the 1860s no one had a clear idea of even the most basic principles of heredity, so the development of an improved strain of plant or breed of animal was by trial and error.

Over thousands of years, many different varieties of domestic animals and cultivated plants have gradually been developed. Once-wild plants changed almost beyond recognition – cereals, for instance, were developed from wild grasses. Cauliflower, broccoli, cabbage and Brussel sprouts all belong to the same family which originated as wild, rather scrawny plants growing on salt marshes and shore-line shingle. Similarly all the modern varieties of chickens developed from the jungle fowl of the Far East. The ancestor of the pig is the wild boar: that of the cow, the wild ox: and that of the over-weight Christmas turkey, the wild and slender flying turkey hunted by the original pilgrims of New England, USA.

The realisation of the importance of Mendel's laws of genetics changed the approach of animal and plant breeders. Indeed, selective breeding today has become part of the science of applied genetics. The process has become more efficient, but it is still recognised that all the answers are not known and that artificial selection is far from being an exact science. Thousands of different traits or features may be involved in breeding experiments and therefore the results are rarely as easy to predict as those of Mendel's seven traits in garden peas (see p.92).

One of the first of the great pioneers of plant breeding was the American, Luther Burbank. He is considered to have been a genius in his work and was responsible for developing a huge list of improved varieties of plants. Perhaps his most famous contribution to agriculture was his success with potato crops. One day in 1871, while searching through a field of potatoes in Massachusetts, Burbank noticed fruits growing on one of the plants. This was unusual because, although potato plants normally have flowers, they seldom bear fruit. New plants tend to be grown from potato stem tubers rather than from seeds. He saved the seeds from the fruits and planted them. He checked the potatoes growing on the underground stems of the resulting plants and saw that they differed from plant to plant. Some were large and others were small; some produced many potatoes, others produced few. One had more potatoes that were also larger and smoother than the others. He selected this plant for future breeding by asexual means. It was named the Burbank variety of potato, which soon became popular throughout the USA.

Burbank's success was an example of breeding from one plant which showed desirable features. Similarly, the first people ever to cultivate plants always saved the seeds for the next planting from the plants that produced the best yield. The offspring are then most likely to have the desired traits and, over countless generations, modern crops have evolved from their wild ancestors.

The same principle can also produce strains of disease-resistant plants. For example, suppose that a fungus has spread throughout a wheat-producing area. Almost all of the wheat has been killed except for two or three plants which survived. Some mutated gene has enabled these plants to resist the fungus. Seeds from these healthy plants are grown the following year. Again the fungus attacks the crop, but this time more plants survive. Over several years this cycle is repeated and each year more plants survive to produce a fungus-resistant strain of wheat.

Closely related strains of the same species can be crossed with the aim of producing offspring showing the best traits of both plants, or hybrid vigour. For example, one parent might be chosen because it grows quickly or resists fungi. The other parent might have particularly good flowers or fruit. Some of the hybrid offspring will hopefully have all the desired traits. Sometimes the new combinations of genes in a hybrid will result in traits not shown in either parent. It may grow larger, have more fruit or resist disease better than either parent.

Engineered plants and animals

Instead of the somewhat hit-and-miss techniques of selective breeding, genetic manipulation can be used to produce ideal crops. The basic techniques for genetically modifying plants are the same as those used for genetically engineering bacteria. Fragments of DNA are taken from the donor organism, introduced into carrier and then put into a new host of cultured plant tissue. There are three main problems encountered with the procedure:

- it is not always possible to pin-point the segment of DNA responsible for a specific desirable characteristic

- suitable carrier organisms are not always easy to find. Many are plant pathogens (microbes which cause disease)

- genes must be controlled when they are in a new host so that they do not spread to wild plants and have any long-lasting unintended effects

The potential for the application of genetic engineering to food production is immense. Insects, fungi and other pests compete with us for our crops. The development of DDT and other pesticides led farmers to think that crop yields could be increased by wiping out these pests chemically. They were mistaken. The bugs evolved to become resistant and, superimposed on this problem were the toxic effects of the pesticides which became concentrated through food chains, causing alarming potential environmental problems. Rachel Carson's best selling *Silent Spring* of the 1960s bore witness to the devastating effects of pesticides on wildlife generally, and on birds of prey in particular.

What if plants could be persuaded to make their own pesticides? This question was asked by innovative geneticists in the 1980s and answered soon afterwards. The first commercial crop plant to be genetically modified was the potato. The insecticide is a toxin produced by a soil bacterium. It is harmless to humans but kills beetles, caterpillars and most other insect pests including the infamous Colorado beetle. Spiders and many beneficial insects are unharmed by the toxin which also breaks down quickly, so it can be considered as environmentally friendly. Geneticists isolated the gene from the bacterium that controls the production of the toxin and inserted it into a potato plant so that the plant could make the toxin in its leaves. When the pests eat the leaves, they die. Of course there is concern that harmful insects may develop resistance to the toxin, and before such a modified plant is used on a large scale, it must receive government approval. In the USA, the Environmental Protection Agency (EPA) has the final say in such decisions. This particular project was reviewed by the EPA's advisory panel in 1995. The window of opportunity was opened for all other crop plants to be modified in a similar way. However, the same window also allowed a whirlwind of opposition to blow through. The repercussions of this have been felt ever since. Fields of genetically modified trial crops have been trashed by those who maintain that they are carers of the environment. In the UK some prosecutions have been made, but in September 2000, those caught in the act of destroying such crops were declared not guilty of criminal damage by a jury. People cannot have it both ways; do they want pesticides sprayed on crops or do they want pest-resistant plants? Unless this choice is made, the starving of the world will continue to die while the affluent will spend their money on slimming aids.

In December 1996, the European Commission approved the sale in Europe of a genetically engineered type of maize which contains genes for herbicide resistance plus a natural insecticide from a soil bacterium. As part of the engineering process, a marker gene was also inserted which marks the cells that contain the modified chromosomes. The marker gives resistance to a commonly used antibiotic. Critics of the technique suggest that there is a small chance that, when cows eat the maize, the gene could pass into bacteria that normally live in the cow's gut. This would help spread antibiotic resistance. Concerns about the presence of the antibiotic-resistance caused the British Government to raise objections to the sale of the modified maize in Europe, but the Commission was forced to consider the financial implications of closing the European market to the product. The USA exports hundreds of millions of dollars' worth of maize to Europe each year and to suddenly prevent this could easily start a trade war.

Towards the end of the year 2000, American consumers learned that they might have eaten corn products (tacos) containing genetically modified maize that had not been approved for human consumption. Amazingly, the American public remained indifferent to the discovery and the most vociferous objections came from the other side of the Atlantic. This observation reveals the developing transatlantic gulf in attitudes to genetically modified foods. Perhaps the American public are more ready to accept the argument that Mankind cannot afford to reject a technology that could deliver cheap food, vaccines and medicines, and help wean the world off fossil fuels.

By 1997, governments of those countries which traded in genetically modified products began tightening up their regulations. The Australia and New Zealand Food Authority (ANZFA) prepared rules which made it much more difficult to sell food made from genetically engineered plants or animals. Approval for the sale of such products is subject to meticulous and prolonged scientific scrutiny. Those foods that are approved must carry a label if they contain more than 5% modified material.

Herbicide resistance

As well as insect pests, weeds compete with us for the crops we plant if left unchecked. For a great many years, farmers have attempted to eliminate weeds by using herbicides. Selective herbicides that kill weeds but not crops are difficult to find so there is great interest in creating a genetically modified herbicide-resistant crop. Progress in this work has been rapid, mainly because the growth-inhibiting properties of some herbicides are the same as occur in certain bacteria. A herbicide-controlling gene can be taken from a bacterium and transferred to a plant such as soya bean via a bacterium.

Again, there have been protests from people who are worried about plants having marker genes which confer resistance to antibiotics such as ampicillin. However, tobacco, tomato, potato and rape seed are crops that have already been modified and are therefore obvious choices for commercial production and attack by carers of the environment.

Besides the problem of the possible spread of antibiotic resistance, what if herbicide-resistant plants escaped and dispersed their pollen, therefore spreading throughout the countryside? How could these 'superweeds' be destroyed if weedkillers cannot affect them? The answer is to genetically modify them, where possible, so that the plants are self-pollinating before the buds open (like the garden pea). In this way, they will not disperse their pollen. Also the seeds can be genetically modified so that they cannot germinate. Their germination is triggered by the production of certain plant hormones. If the genes that regulate these hormones are mutated, germination will be impossible, even if the seeds are dispersed.

Improved products

As well as protecting crops from pests and disease through genetic engineering, geneticists turned their attention to using the new biotechnology to improve the quality of plant products, or to make plants produce entirely new products. One example shows how medical technology has made use of genetically modified plants.

The life-threatening liver disease, hepatitis B, could soon be treated using genetically engineered bananas modified to carry vaccines. Researchers at Cornell University, New York, have produced bananas which are able to manufacture antigens found in the hepatitis B virus. The carrier bacterium is used to transfer the gene for the production of hepatitis B antigen from a virus into the bananas. Theoretically, it could provide a very cheap method of vaccinating populations of developing countries and could be extended to prevention of other viral diseases such as measles, yellow fever, and polio.

Food technology is an important field today in our world of giant supermarkets which have to store foods in bulk. This is another area that is benefiting from gene manipulation. The genetically modified *Flavr Savr* tomato was one of the first fruits to demonstrate the wonders of genetic engineering to the public. Media coverage was widespread when the story broke that scientists had engineered a tomato that could last much longer than usual in a fresh and tasty condition. The scientific explanation sounds much less dramatic than the many newspaper headlines and pseudo-scientific reports that appeared in 1994.

The tomato has a gene inserted which switches off the synthesis of the enzyme which usually causes tomatoes to soften when they ripen. The modified tomatoes are less likely to be damaged when they are harvested and can remain longer on the plant to ripen naturally. They should therefore have an improved flavour as well as a longer shelf life.

Geneticists seem to have an unlimited source of innovative ideas. In the late 1990s, cotton plants were modified so that they produce fibres containing granules of plastic! Although the idea was tried in 1992 using a type of cress plant, it was not until 1996 that the scope for cotton-plastic fibre was realised. This new type of fibre can be used to make fabric for ultrawarm clothes, carpets and insulation. Researchers from an American company inserted two genes from a bacterium into the cotton plant. The bacterium normally makes a biodegradable plastic which is an energy store, rather like fat in animals.

The genetically modified cotton plant can make plastic by a modified natural biochemical process that usually produces oils and waxes in the normal cotton plant. A device called a gene gun was used to fire the two bacterial genes into the cotton plant embryos contained in the seed. First plasmids containing recombinant bacterial DNA were mixed with tiny particles of tungsten. These particles were then stuck on the front of a cylindrical plastic bullet. The device works like a miniature pistol. A firing pin detonates a blank gunpowder charge that propels the bullet down a barrel onto a plate. The impact of the bullet hitting the plate jerks the tungsten and plasmids off the bullet's surface through an opening and across a vacuum to the plant cells. The particles of tungsten are large enough to penetrate cells but they do not destroy the cells.

Built-in fertilisers

Genetic engineering could in future take the place of fertilisers, so avoiding the economic and environmental problems associated with their use. As a rule, plants are adapted to keep microbes out of their tissues. Legumes (plants that produce pods), are exceptions. They welcome invasion by a special type of bacterium which forms swellings on the roots, called root nodules. Both legume and bacteria benefit from the relationship – the microbe has a safe home in the roots of the plant, and obtains sugars which have been made by the plant during photosynthesis. In return, the microbe provides the valuable service of converting nitrogen from the atmosphere into a form that the plant can use to make proteins. Very few microbes have the ability to fix nitrogen in this way. Because of this, leguminous plants like peas, beans, clover and alfalfa do not need nitrogen-containing fertiliser. They have their own built-in bacterial fertiliser factories.

Why does this particular bacterium, *Rhizobium*, live in the roots of legumes and not in cereal crops? It appears that the bacterium responds to chemicals called flavenoids produced by the roots of leguminous plants. Flavenoids enter the bacteria and once inside, trigger the reaction of a key protein which acts as a gene regulator. In the presence of flavenoids the protein switches on certain genes in the bacterium. The bacterium makes a chemical signal which passes through the soil to the plant, telling it to make nodules. This relationship between the bacteria and the legumes is highly specific – a given variety of bacterium can cause nodule formation only in certain groups of legumes. Legumes in the same group make the same type of flavenoid. For example, peas, vetches and lentils all make one type of flavenoid and so belong to the same group. Clovers belong to a separate group, so clovers cannot normally be invaded by the same strain of bacteria as peas.

By 1997 genetic engineers had succeeded in altering the bacteria that normally invade peas so that they will invade clover. The next stage, perhaps, will be to enable the bacteria to invade cereal crops. If the genes responsible for the production of the legume flavenoids could be isolated and inserted into cereal crops, this could be one of the greatest breakthroughs in the history of agriculture. Feeding the world without nitrate fertilisers would reduce protein deficiency in populations and also drastically reduce pollution problems resulting from the use of excess nitrate fertilisers.

A gene too far?

In September 1944, a comment made to Field Marshal Montgomery by one of his lieutenant-generals, before the battle of Arnhem, has become synonymous with being over-optimistic in trying to achieve the impossible. It became world famous as the title of Cornelius Ryan's best seller, *A Bridge Too Far* and begs the question, have geneticists attempted to go a gene too far? Manipulating genes in plants does not usually promote the same ethical concerns as doing the same with animals. We are much more likely to become emotional over a mutated cuddly bunny than we are over a mutated radish! Newspaper headlines like 'Featherless Chickens' and 'Self-shearing Sheep' have a sensational effect.

The idea of monstrosities being purposely made for the sake of scientific research is shocking to the public. Indeed, genetic engineering will provide material for science fiction writers for many years to come. However, let us look in an objective way at some implications of techniques that are already with us.

Life savers

In 1996, Rosie, a cow with a difference was born. Her birth heralded hope for the survival of thousands of premature babies born each year. Scientists genetically engineered Rosie and eight other cows to produce a human protein in their milk. Early in their development they were given human genes which made a protein that is a rich and balanced source of amino acids, essential for newborn babies. The protein could be produced in bulk in Rosie's milk, purified, and added to powdered milk for premature babies. Typically the breed of cow to which Rosie belongs would produce 10,000 litres of milk per year. Even before the success of Rosie, researchers had genetically modified sheep with human genes as early as 1993. These animals were engineered to produce human proteins in their milk to provide a blood-clotting factor needed by haemophiliacs. Another product of a similar technique is a protein which helps treat cystic fibrosis. This is also produced in modified sheep's milk.

The 1990s saw other life-saving applications of genetically engineered animals, including insects. For the first time, in 1996, a mosquito that transmits a deadly disease was turned into a harmless, if irritating, insect by modifying its genes. The disease, in this case is encephalitis in children, which is relatively rare, but if the same technique could be used against mosquitoes that carry malaria, yellow fever, and many other killer diseases, many thousands of lives could be saved.

A fish called 'wanderer'

Most people would appreciate and agree with the life-saving benefits of the above examples of genetically modified animals. Let us now consider some potential problems as a result of genetically changing animals. One well publicised example is the, now ubiquitous, farmed salmon. It is possible that harmful genes will become concentrated in a population bred in captivity, which could escape back into the wild. Fish farmers have designed a 'super salmon' by artificially selecting genes that control desirable qualities like a high growth rate. In doing this, the genes responsible for the salmon's normal homing behaviour have been suppressed. If these salmon escape into the natural wild population and breed, they could introduce mutated genes into the wild population and destroy their normal homing behaviour. The mutants will be doomed to wander the oceans for ever! The results of a study of this problem have indicated that hybrids between farmed salmon and wild salmon survive less well than native fish during the first four months in the wild. This suggests that native salmon have a degree of genetic adaptation to local conditions which is impossible to simulate in farmed fish.

Look – no feathers or wool!

In the 1990s, Israeli scientists were working on producing a featherless chicken. The growth of feathers uses up energy in the animal which could be directed towards making more meat. More meat means more profit. The new technology has hinted at even more bizzare money-making tricks. What if behavioural genes could be altered so that animals become more amenable to intensive factory farming? Such animals would be little more than meat or egg-producing machines with behaviour change so that the animals would hardly be recognisable as such.

Australian geneticists have perhaps demonstrated some dangers of this technology. They have produced a sheep with a genetically engineered skin growth hormone that produces breaks in the wool fibres as they grow. All the wool falls off at the same time so that the sheep do not need to be sheared. Unforeseen side-effects were severe sunburn and spontaneous abortions.

Designer animals

For many years there have been attempts at transplanting non-human organs into humans. The poor success rate has been due to rejection of unmatched tissues, which is a problem even between humans. In an effort to overcome this problem, pigs have been developed that contain human genes. In this century, 'spare part' pigs could be available with kidneys, hearts and lungs ready to be donated to human recipients.

Selective breeding has long been used with farm animals. Since the 1950s artificial insemination techniques have allowed a single bull to father thousands of calves, but their genetic make-up still varies quite widely because of differences in the mother cows. The present quest for genetic perfection had developed a new science of cloning. The result could be a super-race of genetically identical animals, tailor made to produce any protein we want on demand or to produce meat of optimal quality. However, this super-race would be identically vulnerable to the same diseases which could eliminate entire herds very rapidly, as has happened with cloned super-crops.

Mass-production by cloning is well on the way to reality in many countries. In 1997, Australian researchers created almost 500 genetically identical cow embryos. The research leading to the creation of Dolly the sheep, by cloning using a body cell, may lead to hundreds of copies of adult animals being made from a single cell. For example, an élite cow's egg could be fertilised with a prize bull's sperm and then hundreds of genetically élite offspring produced, by cloning and implanting the eggs in surrogate cows.

Will the next step be the cloning of humanised pigs or sheep to supply human blood? This would eliminate the need for human blood donors and provide a bottomless blood bank. Or, perhaps, we have already gone a gene to far!

3 | LIFE WILL NEVER BE THE SAME AGAIN

All life is one

At least two quite different ideas about the origin of matter and life can be traced back to the sixth century BC. One was formulated by the ancient Greek philosophers Thales and Anaximander. Thales believed that the fundamental element of all matter was water. Anaximander, expanding on water's role, is recorded as saying 'Living creatures arose from the moist element as it was evaporated from the sun. Man was like another animal, namely a fish, in the beginning'.

Compared to the relatively tranquil Aegean Sea, the Mesopotanian Plain was an area of violence, ravaged by periodic floods and geological disturbances. This made it easy for the Babylonians and Jews whom they captured to arrive at a different view, called 'catastrophism'.

The biblical stories of the Creation and Noah's flood are regarded today as allegorical by most Christians, Jews and Muslems. But the Old Testament's account of the Creation as a sudden cataclysmic event was accepted literally by the early Christians, and ruled as the prevailing belief in civilized Europe until the mid-nineteenth century.

Yet even during the Dark and Middle Ages, some scholars and philosophers were sceptical and studied the ideas of the ancient Greeks for explanations about the origin of life. They drew particularly on the thoughts of Socretes (470–399 BC), his pupil Plato (438–348 BC) and Plato's pupil, Aristotle (384–322 BC). Aristotle, in his way, was a true scientist. He observed, collected and classified. Memorable is his 'ladder of nature' in which he classified living things in their order of complexity and demonstrated, at least to his own satisfaction, a purposeful progression from lower organisms to humans.

Not until the Renaissance in the sixteenth century, however, did thinkers move toward considering all life to be a single phenomenon. From this concept arose the idea that species could change by a natural process, not necessarily purposeful, and not requiring the special involvement of a Creator.

The new thinking needed bold minds and accurate information. The only references available at the time were rudimentary and incomplete lists of animals and plants. They were poorly described, inaccurately illustrated, confused and muddled by imaginary creatures.

Konrad von Gesner (1516–65) helped set things right with his *History of Animals*, published in Zurich. Perhaps the greatest classifier of all time, however, was Carolus Linnaeus (1707–78) of Sweden. Linnaeus invented the method of naming organisms that science still uses. More importantly, he classified all known living things according to relationships between them; that is, according to their degree of similarity. His work stimulated the development of comparative anatomy and paleontology.

But even Linnaeus himself, did not recognise that all organisms have evolved from a common ancestor. That possibility did occur to the famous French scientist, Jean Baptiste Lamarck (1744–1829), who became the first prominent advocate of organic evolution. It is perhaps an oversimplification to say that he believed the mechanism of evolution to be 'the inheritance of acquired characters'; but he did argue that the effects of use and disuse on an animal's organs could be passed on to its offspring and thus gradually change a species.

Great British naturalists

Proposed by two Englishmen, Charles Robert Darwin and Alfred Russel Wallace, the theory of organic evolution by natural selection shook science to its foundations, revolutionised philosophy and religion, and affected the beliefs of millions throughout the world. It has been said that as a masterpiece of reasoning and exposition, Darwin's *The Origin of Species* is without equal in the history of human thought.

Darwin was 22 when in 1831 he sailed as naturalist aboard *HMS Beagle*. Voyaging around the world, he returned nearly five years later with many specimens, full notebooks and the conviction that forms of life evolved one from another. By the most astute observations, and with his clear mind, he deduced that species could change or evolve over time.

The idea had dawned in South America, where he had unearthed fossils that differed from, and yet resembled, living animals, hinting about the continuity of animal life. Support had come from subsequent observations in the Galapagos Islands, 600 miles west of Equador. There Darwin had found that forms such as tortoises and finches – today called Darwin's finches – differed from island to island, indicating the formation of new species from others already existing. Here was evolution as a result of adaptation.

But what was the mechanism that produced evolution? Wrestling with that question after his return to England, in 1837 Darwin read the new work by Thomas Malthus on survival factors in human populations. Darwin immediately saw that within a given environment, the characteristics of certain individuals would favour them in the struggle for survival.

More and more of them would survive and pass on their characteristics to offspring. Thus all organisms in an environment would increasingly exhibit those characteristics. They would adapt to selective pressures of the environment. This evolutionary mechanism, Darwin called natural selection. In 1842 and 1844, Darwin made abstracts of his theory; but instead of publishing it formally, he continued to seek confirmation in nature.

Natural Selection was not a new concept. James Hutton had written before 1797 that animal forms best adapted to a particular end were 'most certain of remaining'. Unfortunately, his manuscript did not come to light until 1947!

But in 1813, W. C. Wells had grasped the idea that some human characteristics were subject to a selection process. In 1831, Patrick Matthew had published a work on forestry that also recognised the principle. These contributions, however, were not meant to explain the general problem of evolution and Darwin did not know about them until after his own work had been published.

His greatest shock was to come from another quarter. In 1855, Wallace suggested in a paper that one species was descended from another. Thereafter, he and Darwin maintained a long correspondence about evolution. But in 1858, Darwin received from Wallace, who was then in the Malay Archipelago, a copy of an essay that, point-by-point, set forth Darwin's own concept of natural selection!

Like Darwin, Wallace had read Malthus on population and Sir Charles Lyell on geology; and like Darwin, Wallace got the idea of natural selection in a flash – while he was bedridden with an illness. He also had the rare ability of combining original observations with deduction.

Lyell and the eminent botanist, Sir Joseph Hooker persuaded Darwin to present his views, together with Wallace's, at a meeting of the prestigious Linnaean Society in 1858. The following year *The Origin of Species* was published, a book which was to radically change the thoughts of millions throughout the world.

Hard evidence

Fossils are nature's way of writing the history of global biology in the living rock. From the very first, the development of life has been intimately related to changes in the Earth's crust. Moreover, the only records of that development are the fossils accumulated in the rocks that slowly formed over the long ages. Hence palaeontology and geology are sciences closely allied to each other and each is an essential tool of the other.

Yet before the publication of *The Origin of Species*, only a few enlightened sceptics took the view that fossils and rocks told a story of creation, quite different from mythological legends or the biblical account.

The ancient Greeks may have given the first palaeontologist to the world in the person of Anaximander of Miletus, who believed that one form of life might well be the ancestor of another. They may also have produced the first person to link palaeontology and geology together.

In the fifth century BC Xenophanes observed fossilised shells, fishes and seaweeds inland – and concluded that 'a mixture of the earth with the sea is taking place'. These speculations were swamped for centuries by religion-born ideas to the effect that the Earth and all varieties of life on it had been created in an instant by supernatural forces.

Not until the sixteenth century, did thinkers like Leonardo da Vinci clearly see that fossils had once been living creatures and that dry land could once have been the bottom of the sea. Through collection, observation and speculation, scholars of the seventeenth and eighteenth centuries finally created an organised science of palaeontology. By the early nineteenth century, William Smith and Charles Lyell had built up a science that utilised the fossils in rocks to establish the sequence of strata formation and the geological period of each stratum.

Lyell's *Principles of Geology* inspired Darwin's early explorations in South America, and the two men were eventually to become close friends. Lyell had considered that fossils were related not by descent but by likeness of form. He said that were 'abruptly' introduced. Darwin, however, ultimately convinced him that the fossilised species in one rock stratum were descended from other species in the stratum below it.

Darwin gave tremendous impetus to the study of the history of life. To this day, scientists are delving in the earth for clues to the far distant past, testing the theory of evolution by natural selection, checking out its predictions, adjusting its assumptions according to the evidence they find in the rocks. That evidence consists mainly of large numbers of fossils, which may be likened to a huge and growing pile of jigsaw-puzzle pieces.

At first the pieces seemed mismatched, unrelated, as if they belonged to different puzzles and could never be fitted together. Now research has revealed to us that they do actually fit somewhere. But while paleontologists constantly find new pieces, however long research goes on, there will always be some pieces missing.

Yet the missing links do not disprove evolution. They merely demonstrate that some organisms are more likely to become fossils than others. Some of the fossil evidence is remarkably complete, like that tracing the evolution of ammonites, sea urchins, and to some extent, the horse. Other evidence has been forever lost because the animals had no hard shells or skeletons to fossilise. Fossils tell us also that some groups evolve much more slowly than others and that no group evolves at a steady rate throughout the ages. A dominant group may divide in half in as short a time as a million years or even less. Or, like diatoms, lamp shells and certain trees, the group may persist for a hundred million years or more. Evolution, as Darwin was careful to stress, is a painfully slow process proceeding only by the most minute changes from generation to generation.

Finding the plan

Darwin's ideas explained a branch of science called comparative anatomy through common links between structural features of animals. For centuries, people had been dissecting specimens to observe their anatomy but it was Baron Georges Cuvier (1769–1832) who organised such observations so that sensible scientific conclusions could be drawn from them. Cuvier was a notoriously brilliant French scientist who made very important contributions to the study of evolution through his studies of living and extinct species. It was Cuvier who first recognised four basic groups of animals: vertebrates, molluscs, articulates and radiates. Indeed these names were applied to the animal kingdom for well over 100 years after his death. While he did not think that the evidence he found indicated continuous descent in animals, he did relate animals by the form and shape, rather than the use, of particular organs.

When he and other pre-Darwinian thinkers classified their dissected material, they encountered quite a few surprises. Internally, the wings of bats and birds were quite different, even though both groups used them for flying. The forelimbs of air-breathing vertebrates, for instance, showed unexpected internal varieties and similarities. Those limbs are all sorts of shapes and have very different functions in different animals. Some animals use them to swim, some paddle, some run, some fly, others grasp, swing and manipulate. Yet every forelimb contains the same plan of bones; a bone found in one forelimb can usually be identified in all of them. Sometimes the pattern has pieces missing, or two of the bones may have been fused together, but the basic pattern is the same.

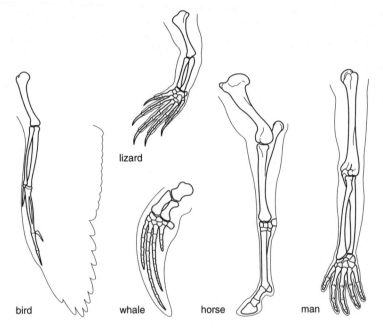

lizard

bird   whale   horse   man

Figure 3.1   Pentadactyl limbs

A fine example of a common plan is seen in the mouthparts of insects. These consist of a central upper lip and three pairs of jaw appendages. In the different major groups of insects, the appendages are adapted to do all sorts of jobs. For instance, the first pair, the mandibles, are large and strong biting parts in cockroaches, sawflies and beetles; but piercing lancets in cicadas, mosquitoes and aphids. The second pair, the maxillae, may be biting tools, sucking tubes or pincers. The third pair are supports, sheaths, sucking tubes or organs of touch. All, however, demonstrate the same basic jaw plan – three pairs of appendages.

The above examples can be explained only by organic evolution. The single plan testifies to common ancestry, with evolutionary divergence occurring as details in structure become modified to fit each insect group to its particular way of life.

In nature, there is also the phenomenon of evolutionary convergence. It is this convergence that produced what, to pre-Darwinians, were completely unexpected internal differences in organs put to the same use. Examples are the wings of birds and bats or the tails of dolphins and fish.

Before Cuvier, scientists placed birds, bats and bees in the same category because all of them had wings. However, the wings of insects are built on a plan which is entirely different from that of birds' wings. The only vertebrates that fly or have flown are the birds, bats and those extinct reptiles, the pterosaurs. The basic vertebrate plan is evident in the skeletons of all three of these groups because they have corresponding, or 'homologous', bones. This, despite each group developing its own wings.

Thus today, the groups are classified as separate and distinct. They become flying vertebrates by separate descent, by convergence. That is, after the air-breathing vertebrates had diverged, three of the branches evolved differently constructed, but successful wings. Though put to the same use, they varied in internal plan.

The way we were

Many flightless insects have tiny remains of wings. Among the amphibians, newts and salamanders, several species have returned to an exclusively aquatic life and have only tiny vestiges of the limbs that their ancestors developed to survive on land.

Some desert-living cacti are plants with leaf-like fleshy stems in which food is manufactured, because this structure saves water. Their leaves, like the newts' and salamanders' limbs are vestigial.

Countless forms of animal and plant life exhibit such vestiges. Indeed, some are not so tiny, like the human appendix or the wings of an ostrich. Why do these ancient remains exist? The answer is that they are the excess baggage of natural selection.

Plants and animals with vestigial organs are descended from plants and animals in which those organs once had a vital function. In some new environment requiring a different mode of survival, those organs became unnecessary and often, a hindrance. So the plant or animal born with even a slightly smaller organ would survive to bear offspring much more often than another born with a slightly larger organ.

Natural selection, in other words, would cause the group to exhibit progressively smaller organs in generation after successive generation. Ultimately those organs might disappear entirely. Where we find a vestigial organ in a living animal or plant, we are glimpsing evolution in action.

A dramatic picture of evolution is afforded us by the development of a complex adult organism from a single fertilised cell. The human egg itself, under the microscope, looks much like a single-celled animal. After fertilisation, when it divides it becomes a two-layered structure with a central cavity that resembles the jellyfish level of organisation. When a mesoderm (see p.89) or third layer of cells forms between the two early layers, the plan of the embryo resembles that of the embryo flatworm – the first group of animals to show primitive sense organs at the front end. Nerve cord and notochord (the forerunner of a backbone) develop next, then gill pouches and a two-chambered heart. At this stage the human embryo looks remarkably similar to a fish embryo. Soon limbs develop and the heart and kidneys change so that the embryo takes on features which make it look like a mammal. However, not until it is almost ready to be born and loses its tail, does it resemble a human baby.

In the words of the eminent English scientist, Sir Julian Huxley:

> *A child of two can tell a pig from a man, a hen from a monkey, an elephant from a snake ... When they are early embryos, they were so alike that not merely the average man but the average biologist would not be able to distinguish among them.*

(see Fig. 3.2 Similarities of embryos).

Does this mean that an animal or plant, in the course of its development, recapitulates the whole of its evolution? Not at all;

but the growth process does present us with many reminders of the various stages of evolution. They tell us unequivocally that there is but one tree of life no matter how numerous its branches. We can be certain that the first multi-celled creature evolved from a single-celled creature and that the first bird, as the eminent Walter Garstang put it in the 1920s, was hatched from a reptiles egg – so the egg really did come before the chicken! The idea of special creation of all things at a single time breaks down at this point with a simple analogy. Why would you build a state-of-the-art formula one racing car via a series of stages of old-fashioned types first?

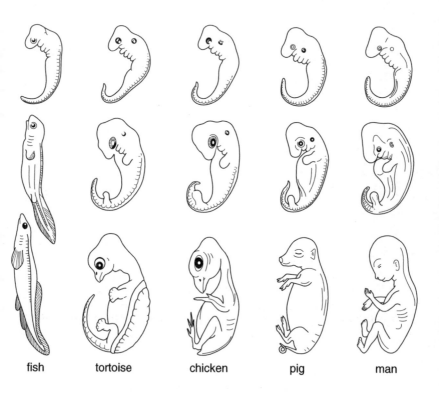

fish   tortoise   chicken   pig   man

Figure 3.2   Similarities of embryos

Where are they now?

Why is the distribution of plants and animals over the Earth not uniform? 'Because the climate and soil are not uniform', you might answer. Yet huge areas of South American, African and Asian tropical forests enjoy very similar climate and soil. Their respective plants and animals, nevertheless, are vastly different. Great orders and families of living things are confined exclusively to one continent or the other.

The explanation is that every group of living things began to evolve from the parent stock not only at a certain time but also in a certain place. By studying the present distribution of members of a group as well as the distribution of their fossil ancestors, we can learn a great deal about the place of origin of the particular group. Consider the present-day members of the camel family. The llama and its relatives, the vicuna and the guanaco are found in South America. The habitats of true camels are in North Africa, Arabia and Asia. They have been distributed over these areas perhaps as much by humans as by nature. While only two types survive, *Lama* and *Camelus*, at times in past ages there were many more. At least 25 fossil types have been identified. Originating about 50 million years ago, all but one of those types lived in North America. The only fossil camels found outside North America date from two million years ago and belong to our two living types or to *Paleolama*, an extinct form of South American llama.

In spite of the modern distribution of the camel family, then, clearly its origin of evolution was North America. It is equally clear that camels branched off from other cloven-footed mammals at least 50 million years ago, and that they did not spread to other parts of the world until toward the end of the Ice Age a million or so years ago. Relative newcomers to the regions where they are now found, these beasts are now entirely extinct in the lands where their ancestors roamed.

Marsupials (pouched animals) also sharply define a place of evolutionary origin, or 'theatre of evolution'. The evidence indicates that they originated on the South American and Australian continents, which were connected at the time. Australia separated from all other land masses before more advanced mammals gained a foothold. Today, all mammalian life 'down under' is either egg-laying, like the platypus, or marsupial. Other, more advanced mammals have been imported by humans.

Theatres of evolution need not be of full continental size. However, for a family of animals or plants to spread from one great theatre to another, long ages of time are required and communicating pathways must exist. The emigrating group must somehow cope with water barriers, high mountain ranges, or zones of hostile climate. It must locate bridges, entrances and enough suitable food to survive.

The story of geographical distribution has only begun to unfold. Palaeontologists and naturalists continue to seek clues to reconstruct the travels of plant and animal life which have taken place over hundreds of millions of years. We are now sure that the human theatre of evolution was in Africa. Much remains to be learned about theatres of evolution in the oceans.

What the known facts about geographical distribution tell us so far, is that a form of life originating in one area may emigrate to another. There, as natural selection acts upon it, generation after generation, that form may adapt to conditions in the new area and survive; but on separate parts of the Earth. Though their climate and other characteristics are alike, similar life forms need not necessarily be produced.

A change is as good as a mutation

No one can predict accurately what will happen in the evolution of living organisms. The future progeny of a seed-eating bird may evolve into an insect-eating bird if insects are available in great numbers over a long period of time. If other insect-eating birds are present to compete with this food, however, the seed-eaters may starve and become extinct. Indeed, the climate may also be a factor which will determine the survival of a species. Any change may be beneficial to one species but not to another. Survival of any organism depends on chance. A slight change in climate, food or even a single gene mutation may spell death or enhance survival. The long history of evolutionary development through adaptation is a web of threads woven by chance.

No two living things are alike unless they have come from the same egg and have identical genes. If you look casually at a group of animals of one species, you might think that they were all alike – or at least that the adult males and the adult females were. But anyone who has kept and cared for animals soon learns to distinguish them. Plants can vary just as much as animals.

A boy tends to resemble his brother or his father more closely than more distant relatives, and he resembles relatives more closely than his unrelated friends. These resemblances are due mainly to heredity, and the differences that enable us to tell one twin from another are due to the environment.

Separation of chromosomes in meiosis (see p.96) and the random way in which they recombine also produces variation. Even the simplest of organisms have hundreds of genes and complex organisms have tens of thousands. A tremendous amount of variation is possible from this number alone. Crossing-over during meiosis increases the possibility of variation by separating linked genes and linking genes in new combinations. The mixing of chromosomes at fertilisation ensures that all individuals are unique. Other factors affect development as well as the genes that the individual inherits from its parents. A pea plant may inherit tall genes, but it will not grow tall in poor soil. Only if it has a gene for shortness will it not grow tall however good the soil. Genes determine what an individual *can* become, but environment, together with the genes, determine what it becomes. Thus, we can see that the variation of living things is due both to heredity and environment.

Regardless of how many times different genes are combined in new arrangements, the results are all variations on an existing pattern. But a change in the genes themselves 'a mutation' will give a new pattern.

One of the first mutations to be studied occurred in the fruit fly, *Drosophila*. In the 1930s, Thomas Hunt Morgan, a famous American geneticist, discovered a wealth of information with his studies of the genetics of this unlikely animal. One day he found that among a strain of pure red-eyed flies there was one with white eyes. This change was caused by a sudden and unpredicted change in one of the many eye-colour genes. Since then, hundreds of fruit fly mutations have been found and studied in detail.

Mutations can be caused by a change in the structure of a gene, in the arrangement of genes on a chromosome, or in abnormalities in numbers of chromosomes in a cell. Radioactive materials and certain chemicals can alter the structure of DNA and hence the structure of a gene. Since genes control the manufacture of a protein, the result of a mutated gene may be the inability to produce one or more essential proteins, leading to abnormalities or death.

Most mutations are recessive and are masked by the dominant normal genes. Thus, except for certain sex-linked genes (see p.100), they do not show up until two of the mutant genes occur in one individual. Some mutations produce only minor changes in body chemistry. If it is an unfavourable change for the organism it will be lost, because the organism with it is unlikely to reproduce to carry it to the next generation. Sooner or later, however, the mutation will occur again.

Though we speak of mutant genes and normal genes, the genes we now call normal were once mutants, but because they were favourable, they have become part of the normal collection of genes. This is all part of the process of evolution.

Darwin's legacy

We can now develop the idea of natural selection in the way that Charles Darwin himself worked it out, with three facts and two deductions from them.

The first fact

This was stated first in 1798 by Thomas Malthus, whose *Essay on Population* was read by both Darwin and Wallace. Malthus wrote 'Population, when unchecked, increases in a geometric ratio'. Animals and plants have a tendency to multiply at a geometrical rate in numbers that run, 2, 4, 8, 16, 32, and so on. The offspring always tend to be more numerous than parents. We only have to look at the human 'population explosion' to see this in action.

The second fact

While all living things can increase at a geometrical rate, they seldom do. Few species, apart from humans and some animals and plants that depend on us, have been observed to increase so rapidly over a very long period of time. The species that have done so have often been presented with quite new opportunities by an accident of nature or by being introduced to a new suitable habitat by humans. Examples include the European rabbit in Australia and certain species of gulls on rubbish dumps. Both of these have undergone population explosions.

The first deduction

This can be called 'the struggle for existence', or, more accurately, the competition for the chance to reproduce. Almost everywhere in nature, animals and plants produce more young than can possibly survive to reproductive age. They must compete for food and for other needs of survival before then can reproduce. One only has to observe the common housefly to realise that this sort of competition prevents us being completely covered with a great mass of flies even though they breed at an alarming rate. Most animals in nature do not survive long enough to breed.

The third fact

All living things vary.

The second deduction

This is what Darwin called 'natural selection'. Simply put, natural selection states that the competition is for existence between individuals which vary among themselves. Thus some individuals must be more likely to succeed than others. Those with favourable variations will be more likely to survive and reproduce themselves than those with unfavourable variations.

As we have seen, a great deal of variation is inherited. Favourable inheritable variations have a better chance of being incorporated into the next generation than unfavourable variations.

Natural selection is the principal agent of evolution. It is not a physical force; it has no purpose; it is a process that occurs completely by chance, like evolution itself. It is a process that has turned simple elements in the environment into material of wonderful complexity, and has made, through millions of generations, the human brain, the bird's wing and the bat's ear, all from primeval protoplasm.

A question of species

A species can be defined as a natural population of any type of organism that breeds within the group but does not breed with other groups. The principal cause of species formation is geographical isolation. For example, if a species of warm-blooded animals lives in several latitudes, the ones nearer the poles may be larger than those nearer the equator. There is a good explanation for this. Larger animals lose heat more slowly than smaller ones of the same kind.

Within each species, the various characteristics of the population show gradations through its range. Sometimes these variations are continuous; more often they appear as a series, with definite breaks associated with a population in a definite region. We can call these populations subspecies.

Geographical subspecies can sometimes become isolated, so that they may continue to evolve on their own. Sometimes this evolution in isolation is so successful that the subspecies begins to spread. Eventually, some individuals will, at the edge of a new range, begin to come in contact with those of another population. This is the testing time. Sometimes the separate evolution of the two populations will not have gone far enough to prevent interbreeding or hybridisation between them. After a period they will settle down together, still members of the same species.

But sometimes, the two groups will settle down as similar, but truly separate species. Even if they can still breed together, they will not normally do so. As time goes by, they will tend to differ more and more.

This geographic speciation is typical of nearly all animals, and seems to be the only way species are formed among birds. Processes are known among plants and some lower animals which can result in instantaneous production of new species in one place. These processes are connected with an increase in the number of chromosome pairs beyond their normal number, and is known as polyploidy.

Adaptation

The places and conditions in which an organism lives are collectively called its environment and it is impossible to think of life without thinking also of the conditions surrounding it.

For a species to survive, its members must be suited to their environment in such a way that they obtain enough food, escape from their enemies and are able to protect themselves from climatic change. The main requirement is that they must survive to reproduce and guarantee the survival of their offspring. The adjustment of life to the environment is known as adaptation, and in the sense that all species that exist today are the result of 4,000 million years of evolution, all of them have adapted to the conditions of their present surroundings.

Evolution has produced an amazingly varied array of organisms, wonderfully adapted to every corner of the Earth, with organs and types of behaviour so intricate and beautiful that it seems amazing that they ever developed.

With the perspective of evolution's eye, we can begin to understand how plants and animals came to be what they are. We can spot the different kinds of adaptations in the same animal or plant family or order; or the same kinds of adaptations in completely different groups. We can learn, for instance, that evolution can make a mole out of an advanced insect-eating mammal in Eurasia and the United States, and also out of a quite unrelated marsupial in Australia. Whenever we think about the tremendously varied forms of life on our planet, we must keep its long history of change over thousands of millions of years in mind.

4 | LIFE'S RICH TAPESTRY

Living in harmony

If the ceiling of the Vatican's Sistine Chapel was not so large, Michelangelo would have had some serious marketing problems. When viewed from close range, his masterpiece seems like a random collection of figures in all sorts of poses and in various states of undress. From a distance, however, and notwithstanding an aching neck, the individuals are seen to be organised into one of the most magnificent works of art ever produced in the history of the world. So it is with the organisation of life on our planet.

Almost any part seems to stand alone, but when one steps back, the great panorama of interrelations becomes apparent. Before the large scenario could be seen though, the smaller parts had to come into focus. Specialists provide us with knowledge of the individual parts; generalists, called ecologists aim to gain knowledge of the whole picture. So ecology then, is the study of the interrelationships of organisms and the environment. In short, it is the study of the grand scale of life. Its title was coined by Ernst Heinrich Haeckel (1834–1919) in 1866. Haeckel was a well known German field naturalist and became Professor of Comparative Anatomy in Jena. He developed the word, ecology, from the Greek *oikos*, meaning house and *logos*, meaning treatise or 'study of'. Thus, its beginning was the study of where something lives.

Simply put (perhaps too simply!), an organism's habitat is its address and its niche is its occupation, but academic purists would need greater accuracy when defining these terms.

A habitat is the place where an organism makes a living naturally and can be described in several ways. For example, an animal may live on a rocky shore, or more specifically, in a rock pool. Furthermore, it may live in a certain part of that pool, called a microhabitat. Wherever an organism lives it interacts with its surroundings in a multitude of ways. The principle is the same as the old worn out cliché 'no man is an island'. The effects of the surroundings (the environment) influence the organism which, in turn, interacts with other living things around it. The sum of all such interactions, along with the organism's own requirements, describe its niche.

The word, niche, is derived from the Italian, *nicchia*, which means a cavity into which something fits. In ecology it is the total functional role of an organism in a community. The niche encompasses all the links between the population, the community, and ecosystem in which an organism is found.

Two species cannot occupy exactly the same niche indefinitely. If two species were to find themselves in such a situation, it is generally predicted that one would dominate the other and the victor would eventually replace the vanquished. Where species do coexist, then, we can assume that they are interacting with the environment in different ways (occupying different niches). This means that when you see various species of seed-eating birds in the same wood, they are utilising the habitat differently. Sometimes several species of the same type of birds can survive on food provided by the same tree. They seem to be occupying the same niche but they actually divide the tree into different feeding zones and each species confines its feeding activity to a particular zone. By exploiting different parts of the tree, the species reduce their competition for food and thus they can occupy the same habitat.

Animals may also divide up a habitat in other ways. For example, they may utilise the resources at different times in a 24-hour cycle. Some are 'night shift' workers, and others feed during the day. Others use resources at different times of the year.

View from a height

We tend to think of our planet as always being there to unfailingly provide us with all things necessary for life. In short, many of us take Earth for granted. All too often we forget that life exists only in a thin film covering the surface of an immense ball – something like the skin on an apple. Within the skin of the apple, sunlight and water react to help ripen and mature the fruit, just like they have done to permit life to evolve and mature on Earth. The fragile film is the biosphere and is responsive to a number of influences and, hence, is highly variable from one place to another. Also, each place is likely to be unstable, so that it changes with time. Nevertheless, the different kinds of 'places' in which life exists on land can be roughly categorised based on physical and biological properties. The 'places' are called biomes and defined according to the plants they support. Of course, the make up of the plant community is dependent on factors such as soil conditions, available water, weather, day length, and competition. However, two climatic factors, water and temperature are the main determinants. Certain kinds of animals occupy each type of biome, because different species of animals are dependent on different sorts of plant communities for food and shelter. The major biomes on Earth are:

- **Temperate deciduous forests** where the dominant trees are hardwoods and which once covered all of Central Europe and the eastern United States. This biome is subject to harsh winters when trees shed their leaves so that water loss is minimised. At such times most water will be locked away as ice in the upper layers of the soil. The trees experience warm summers that mark periods of rapid growth and rejuvenation. Before new leaves begin to shade the forest floor in the spring, a variety of herbaceous plants may appear. These are annuals and, therefore by definition, complete their life cycles in one year.

- **Grasslands** where the soil is porous and which support species of grass ranging from all those found on prairies to the massively thick-stemmed bamboos of the tropics. With erratic rainfall to cope with, any trees growing in such areas typically line rivers. Grasslands may be found in the tropics or in temperate zones. The large gregarious herds of grazing herbivores are supported in such areas like the African savannah.

- **Deserts** with the extremes of very cold nights and the oppressively hot days. The sparse rainfall comes in sudden downpours, so that much of it evaporates or runs off, sometimes causing flash floods and erosion. Directly after the rain, annual plants take advantage of the transient moisture and explode in an orgy of blossom so that they can produce seeds. Other desert plants meet the water-shortage problem in other ways, either by storing water in modified parts of their bodies or by cutting water loss by reducing the surface area of leaves and having thick weather-proof coverings. All desert animals must beat the heat and conserve water. They do this via behavioural modifications like being nocturnal and by physiological adaptations to their excretory systems which allow them to recycle the limited water that is available in their food.

■ **Tropical rain forests,** defined by torrential rains that fall almost daily during the summer. They are found mainly in the Amazon and Congo Basins and in Southeast Asia where the temperature doesn't vary very much throughout the year. A great diversity of plants and animals are found in this biome. Trees grow to a tremendous height with their branches forming a canopy. The floor of the forest provides a dark and steamy habitat for myriads of insects and birds, but is quite open and easy to traverse, in contrast to jungles (see below). Animals may breed throughout the year because of a continuous abundance of food and competition is generally considered to be very keen due to the enormous biodiversity. Where sunlight is a able to penetrate the leaf canopy, jungles are formed. These are the densely vegetated and tangled back drops for the plethora of 'Tarzan'-type movies, familiar since films were invented. A century before, however, many explorers of the unknown tropics gained a mistaken impression of the density of vegetation of true jungles because they covered large distances on water. Perhaps understandably, they preferred to pit their wits against the dangers of the rivers rather than chance their luck by hacking their way through the land-bound dense foliage with its seething and swarming animal life.

Their descriptions of their travels coloured the literature for many years and also confused the terms, tropical rain forests and jungles.

■ **Tundra** which is covered throughout most of the year with ice and snow is the prevailing biome of the far north (arctic tundra), but it may also appear on the peaks of mountain regions such as the alpine tundra and on the Rocky Mountains. It enjoys summer for no more than two to four months, just long enough to thaw out about a metre of soil above the permafrost (permanently frozen soil). The hardy plants of such a biome, like dwarf willows and birches, disguise its fragility because, once disturbed, these areas take very long periods to recover. Animal life on the tundra is surprisingly abundant for such an apparently inhospitable place, but it is confined to the warm blooded types.

■ **Taiga** which is confined almost exclusively to the Northern Hemisphere and is identified by vast areas of coniferous forests of pine, spruce and fir. Some of these trees are the largest living things on Earth and include the giant redwoods of North America. Winters are long, cold, and wet, while the summers are short growing seasons. Extensive bogs occur and the forest is covered by a carpet of pine needles which allow little growth of competing plants. The dim light allows some mosses, ferns and a few flowering plants to survive and the animal biodiversity is relatively small and limited to some birds with specialist feeding habits. Mammals include moose, bear, rodents and wolverines.

Water world

Despite the fact that most of our planet is covered with fresh or salt water, and therefore it constitutes a very important part of the biosphere, we often neglect it. Apart from drinking or washing, we generally avoid it unless we need to make a living from it, or use it for recreation. If we stay in it too long, at worst we drown, and at best it clogs our ears, wrinkles our skin, is rarely at a comfortable temperature, and is often hard to see through. Thus we probably imagine that we know more about its role in ecology than we really do. Yet our lives depend on it and we are usually cheered by the sight of a clear mountain stream cascading as a waterfall or by the gentle embrace of waves on a shore.

Ecologically speaking, a neat and over-simplified classification of the world's water puts it into the categories of salt and fresh. This labelling, of course, breaks down in estuaries, brackish tidal saltmarshes, and mangrove swamps. Here, salt mingles with fresh water and provides a habitat for very specialised plants and animals. They have to be able to withstand the rigours of both worlds and this puts enormous demands on the ways in which they balance their salt content. By definition, fresh water should have no more than 0.1% salt, whereas proper sea water has about 3.5% salt. Of course, varying temperature will vary the concentration of salt because of evaporation and an increase in precipitation will cause dilution.

Lakes and rivers provide less stable habitats than do oceans of seawater because of their smaller size. For example, some lakes may evaporate to a fraction of their former volumes, thereby concentrating their soluble contents. By mixing with the material on their bottoms, they may then become muddied and opaque. There relatively small volumes allow them to become polluted rather easily and their temperature may fluctuate widely. The smaller the body of water, the more it is at the mercy of outside influences.

Due to the peculiar vulnerability of bodies of fresh water, human activity has all too often altered them drastically. The human race has perfected the art of polluting the environment on a world-wide scale. Despite recent attempts to redress the balance, there are still significant areas of fresh water which can no longer support their former biodiversity. All nations which have lakes and rivers to provide natural resources have managed to affect their potential productivity by various forms of pollution. The shame is that, in many cases, the damage is irreversible.

How to cause a disaster without really trying

One reason why our species has become top of the league table of polluters is because of success in a process called eutrophication ('good food'). This is a normal aging process of lakes but the use of detergents and chemical fertilizers has increased the amount of phosphate and nitrate that are washed into the water. These join other forms of human waste in our fresh waters and speed up the aging process and accelerate the production of organic matter in lakes. With increased plant growth, there is a corresponding increase in death rate and the dead plants sink to the bottom, eventually making the lake shallower. So much oxygen is used by the microbes that cause the dead material to decay, that the water becomes devoid of oxygen. In time, species that need most oxygen die out completely and are replaced by those few that can survive in an oxygen-poor habitat. A once fish-packed lake is thus changed into an aquatic desert.

Sea world

Oceans cover three-quarters of the surface of the Earth and if smoothed out, the whole planet would be covered entirely by water. This is, indeed, a watery planet. Sailors, beachcombers, poets, and all with a degree of romanticism in them, are invariably attracted to the sea, enjoying its compelling mystery in their own ways. Perhaps we are exceeding necessary technical descriptions at this point, but the wonders of ecology are more than just technical. However, ecologists, like other scientists, need the direction of definition and description.

The average depth of the oceans is about three miles, but there are places where the water is seven miles deep. So the deepest oceans are deeper than the tallest mountains are high. Prophets have predicted that the oceans will, one day, provide us with most of our food. Simple maths will show that this is not really feasible. We presently take only 3–5% of our food from the seas, and even if we were to double that in the next decade, our food problems would not be solved. This is because with the world's population increasing at its present rate, doubling the produce from the sea would add only 3–5% more to a population that has increased over 18%. Superimposed on this is the complication that the deep oceans are just like deserts and are not very productive. In fact many parts of oceans are almost completely devoid of life.

Where currents are deflected upward by the mountains on the ocean floor, cold water from the ocean depths wells up to the warmer surface. This cold water carries with it tonnes of accumulated sediment from the floor. The sediment is natural fertilizer and promotes the bloom of life in the surface layers of the oceans. Chlorophyll-bearing plankton, phytoplankton, have an easy life at such times. These primitive plants simply drift in the light-filled, warm, nutrient-rich oceanic soup and spend their time photosynthesising and reproducing. As a result, animal plankton flourish on their ample food supply. These tiny organisms, barely visible without a microscope, form the basis of the oceans' complex food webs. One litre of sea water may contain 12 million phytoplankton. These food producers, however, can live only near the oceans' surface, since the sunlight needed for their photosynthesis cannot penetrate below about 200 metres.

Not only are phytoplankton essential as food producers, but they also release much of the Earth's oxygen as a waste product of photosynthesis. Thus the importance of maintaining viable oceans is apparent. All nations recognise the fragility of the food webs in the sea and express their desire to minimise pollution which is the main cause of their destruction. The same nations, however, often rely on oil to sustain their economy and huge quantities of this fuel are transported by ever-bigger tankers through environmentally sensitive areas. Concerns about conservation invariably take a back seat in discussions relating to deals worth billions of dollars in our oil-dependent world. Frowns of consternation appear on the brows of all of us when the media carry stories of oil spillage disasters, but memories are short and after the initial outcry at seeing pictures of horribly oiled animals, we quickly forget as we fill up at a petrol station and then drive away from the problem in our cars.

On Good Friday, March 24, 1989, the skipper of the *Exxon Valdez* allegedly steered his massive oil tanker on an incorrect course and retired below deck. At 12.04 a.m., the ship smashed into the well known and chartered Bligh Reef in Alaska. More than 10 million gallons of crude oil vomited into the previously unspoiled waters of Prince William Sound. The resulting devastating lethal sludge soon covered a distance equal to the whole coastline of California and ruined many complex food webs besides directly killing innumerable birds and mammals. No one can be certain that the area will ever recover completely. This is just one example of disasters involving the spillage of oil which occur almost annually, somewhere in the world.

Life on the bottom

People who have descended to the depths of oceans in pressurized bathyscapes have reported that they can detect light at depths as great as 600 metres. The light that reaches these depths is pale blue, since the red and orange of the light spectrum have been filtered out by the water above. Many of the fish that live there are reddish in colour when they are viewed in ordinary light. In their natural habitat, they appear dark and shadowy, since there is no red light for their skin to reflect. Their red colour absorbs what blue light there is.

It was once believed that nothing could live below 600 metres because the pressure there was too great for any form of life to withstand it. Then, in 1858, a cable was hauled up from a depth of 2000 metres in the Mediterranean. Marine biologists were astonished to see that it was covered with all sorts of living encrustaceans, so some serious thinking had to be done about the effects of pressure on living things. Now it is thought that living things can survive at any depth as long as they are able to develop pressure inside their bodies equal to the pressure outside. This pressure can be so great that fish brought up alive from depths exceeding 600 metres have literally exploded on reaching the surface!

The deepest parts of the oceans are only now beginning to be explored in detail. They have been the source of astounding discoveries which turned some classic theories of interdependence on their heads. It was not so long ago that students were taught by their professors of biology that all life depended on photosynthesis and, therefore on light. The communities of so-called abyssal life depend on certain bacteria which can change the chemical energy in minerals into the chemical energy that they need themselves. Thus they can be at the base of a food chain as producers, without using sunlight as the initial energy source. The communities keep at a suitable temperature by water being heated by the Earth's core and escaping through vents. An upwelling of minerals at these vents also occurs and the end product is a complexity of food webs with bizzare collections of animals including molluscs, crabs and tube worms.

Where water meets land

The land-ocean interface that we call the sea shore has been the source of so many works of art and literature all over the world, but so few realise the full impact of this region on their lives. The importance of the coastal ecosystems of the world lies in the fact that life is much more abundant there than in any other part of the marine environment. The edges of the continents extend up to 150 miles beyond the shore as continental shelves. At these relatively shallow depths, sunlight easily reaches any plant life and allows it to flourish. In turn, the plants provide food and shelter for animals. Each part of a coastline has its own characteristic features which include rock, sand or mud in various combinations. In the warm tropical oceans, coral reefs play an important ecological role with the greatest biodiversity found in the sea.

Rocky coasts are often the most awesome navigation hazards for seafarers, having dangerous jagged formations and hidden, submerged reefs formed by constantly pounding waves. Such coastal areas boast a remarkable array of plant and animal species found in distinct zones, determined by tidal range and exposure to wave action. Certain species are confined to their own zones, so there will be some molluscs and algae at the top of the shore which can withstand more desiccation than those on the bottom. Also, some species are better adapted to living on very exposed shores than those that live on sheltered shores.

Sandy shores have a relatively smaller biodiversity than rocky shores, which is perhaps to the advantage of timid bathers who want to avoid the feel of the odd, mysterious 'thing' on their feet. The reason for the apparent scarcity of life is that wave action causes constant shifting of the sandy bottom, depriving many species of a fixed surface to which they can attach. Any animal that lives in such shifting sand must be able to scurry along very quickly, like crabs, or bury themselves in haste, like cockles and some clams.

Unfortunately, the weaver fish and stingrays do both! Sand dwellers that live in areas which have large tidal ranges, such as in the Bristol

Channel, encounter exceedingly severe conditions at low tide when they are exposed to the parching sun and predation by birds.

Similar to sand, but less aesthetically pleasing is mud. Mud flats are probably the least appealing part of the seashore, and this perhaps explains the general lack of interest in protecting them. They can rarely be described as beautiful, are poor places to spread a towel, and they have their own peculiar smell. A cursory glance at low tide will not provide the sight of much life, especially if you are noisy and have frightened the feeding birds away. In fact, mud flats may harbour tonnes of hidden animals. They are submerged at high tide and exposed at low tide. Snails, crabs, shrimps, worms, fish, birds, and a multitude of invertebrates with no common name, make a good living on or under the mud. Each square metre of mud flat can support thousands of invertebrates by allowing a film of growth of single-celled plants call diatoms as producers in the food web. Many juvenile marine animals begin their lives in nurseries on mud flats, only later moving out to take their place in the mysterious pageant of the open sea. Hence, the life of the ocean itself is dependent on the preservation of mud flats that may seem unsightly and repugnant to many. Unfortunately, mud flats and salt marshes have attracted the eyes of entrepreneurs and commercial developers who are aware of the public's lack of interest in them. Waterfront housing developments, industrial sites, leisure-based marinas and barrages have produced stretches of paved, neon-lit promenades which all attract the universal god of 'cash' in one form or another. Vast stretches of the coastlines of industrialised nations have been treated this way – all in the name of progress. Where once the curlew piped its plaintive cry, someone is treading on the ubiquitous splodge of chewing gum on a paving stone or even something worse if our canine 'best friends' have been there first!

Now that we have described some of the areas and habitats of our planet, let us consider some of the ways species interact.

Relationships

We will begin a discussion of interrelations by defining some essential concepts. We have seen that biomes are usually huge complex masses of land and are not always precisely identifiable because they overlap. However, they are composed of smaller components that can be more accurately measured. For example, biomes can be broken down into ecosystems. Typically, an ecosystem is a group of interacting living things, together with all the environmental factors with which they interact. An ecosystem is considered as an independent unit with light as its only outside energy source. Conceptually, an ecosystem may be a pond, wooded area, or a field. Each of these is too large a unit to study in detail, so they too must be broken down to smaller units. One such unit is the community which is an identifiable interacting group of populations of different species. Whereas ecosystems include abiotic or non-living factors, a community includes only biotic or living factors. A population can be described as interbreeding groups of organisms.

Reaching a climax

If you were able to watch specific areas of land over the world for a long period of time, you would notice that in some places communities of life changed. Ecologists call these changes succession and define it as an orderly sequence of species, structure and energy flow in a specific area over time. Those areas that are not changing may have gone through successional stages and have now reached a climax stage, where the species, structure, and energy flow will remain relatively stable as long as the environment does not change markedly. Biomes are defined by their climax vegetation.

There are two fundamental kinds of succession; primary, and secondary. Primary succession involves the early progression of species where no community previously existed. In other words, primary succession starts from scratch. Although we find primary succession exhibited after the introduction of life to rocky outcrops, river deltas and sand dunes, the most spectacular example is seen after volcanos. Among the most dramatic in recent times was the recovery of the devastation caused by the unexpected eruption of the volcanic Mount St Helens in 1984. Many ecologists were surprised at seeing how rapidly the ravaged landscape began its recovery after being so overwhelmed by the titanic forces of nature. Within a year, flowering plants were blooming on the volcanic ash. Primary succession usually begins when small, hardy, drought-resistant species called pioneer organisms, invade the lifeless area. On rocky outcrops, for example, lichens may be the first to get a grip on the substrate, held fast by their tenacious, water-seeking fungal component, while the algal component photosynthesises to make food to share with the fungus. Lichens act like crow bars, forcing open the minutest cracks in rocks and allowing debris from dead and decaying organisms to accumulate. The decomposing bodies add nutrients to the microhabitat and soon the rocks become home to mosses, ferns and then grasses as they follow the pioneers, one after the other, providing living scenery for the drama of life that follows.

As plant roots penetrate the crevices, they exert a remarkably powerful pressure in the cracks and continually widen them. Populations of animals which help decomposition flourish by feeding on the dead and decaying material (detritus). The lichens, which made it all possible in the first place, eventually succumb to the pressure of competition and their presence is masked by a steadily increasing diversity of other organisms.

Secondary succession is the sequence of communities as ecosystems recover in disturbed areas where once soil supported life. The rate of community change can be much more rapid than in primary succession as a consequence of suitable soil already being present. The classic illustration of secondary succession is all too often seen on derelict and abandoned farms. The first colonisers are grasses, then shrubs and fast growing softwood trees become established. As they grow, their branches begin to shade the soil and shade-tolerant weeds replace the initial invaders. Following these may be slower growing and stronger hardwoods which allow a climax stage to occur. Secondary succession can be relatively brief in some cases and very extended in others. In grasslands for example, a climax can appear on cleared land after only 20 to 40 years, but in the fragile Arctic tundra, recovery can take centuries.

A challenge to a weaver

Who was Arachne? Arachne was a Lydian maiden of ancient Greek mythology, so skilful at the art of weaving that she challenged the Greek goddess, Athene, to a weaving contest. Athene lost the contest and was so jealous that she tore Arachne's work to pieces.

Arachne was so despondent that she hanged herself. Athene intervened and cut the rope, thus saving Arachne's life, but she changed Arachne into a spider and the rope into a spider's web. Arachne was destined to become the finest spinner of webs in the ancient Greek universe – if you believe that, you will believe anything! However, today, we remember this charming legend by giving the name, Arachnida, to the animal group to which spiders belong. The challenge would be for Arachne or anyone else to weave a web as intricate as the web of life on Earth.

Life on our planet is tied together in a complex web in which energy is exchanged between those who eat and those who are eaten. A minute percentage of the sun's energy is captured by producers but it is enough for them to convert into energy stored in food by photosynthesis. From the producers, which are always plants, a proportion of the original energy from the sun is passed to various consumers. These are organisms that must rely on others for the organic compounds that they require.

Since some consumers eat producers, others eat other consumers, and some eat both, the flow of energy through consumers involves several feeding (trophic) levels. Between each level, energy is lost to the environment as heat from respiration and as excretory products. Thus we have primary consumers, the herbivores that feed on producers; secondary consumers, carnivores that feed on primary consumers; and so through to tertiary and quaternary consumers. If there were more trophic levels than this in a food chain, it would be a rarity because as the chain proceeds, so more and more energy leaks away and there is simply not enough to support any more.

View from the top of a food chain

We live at the end of food chains as this simple example will show:

rice ⟶ grasshopper ⟶ frog ⟶ trout ⟶ human

Now if humans were to eat trout alone, we would have to catch one nearly every day, and about three hundred of them would be needed to support each of us for a year. Each trout would eat a frog every day. Therefore 300 trout required to support one of us for a year would, themselves, eat 90,000 frogs annually.

Each frog would eat a grasshopper every day, so the frog population would consume 27 million grasshoppers in a year. A herd of grasshoppers that big would eat something like 900 tonnes of rice. The large numbers of organisms at the beginning of the food chain are needed because none of the organisms is very efficient at converting food into body tissue. Only about 1–2% of the sun's energy striking the paddy field is converted to rice. The grasshoppers are able to convert only about 10% of the rice into grasshopper; most of the rice is uneaten, undigested, or used to provide energy for what grasshoppers spend their days doing – chewing, hopping, and reproducing.

Likewise, humans, trout, and frogs are able to convert only about 10% of their food into human, trout, and frog tissue respectively. It is apparent that the number of animals that can be supported at the end of the food chain is directly related to the number of links that the chain has.

We can support more people on the food chain by shortening it. Get rid of the trout and the paddy field will produce 90,000 frogs for human consumption. If each person could get by on 10 frogs per day, the frogs would support 30 people for a year. But who likes frogs? Let's go to the grasshoppers and assume that 100 a day would satisfy you. (I imagine that you might get along on fewer than this on the first day!) It is now clear that 900 people could live on 27 million grasshoppers.

If we eliminate that last animal link in the food chain and eat rice, the paddy field will support even more people. About 2,000 people, each eating a couple of kilos of rice per day, could live on the area of paddy field that supported only one fisherman!

This simple logic requires the vast populations of China and India to eat mainly rice, or at least mainly plants. However, people in Western Europe, USA, Canada, Australasia and in most western societies enjoy a varied diet that includes large amounts of poultry, beef, lamb, pork and fish. The constantly rising price of meat in the shops shows us that our expanding population is pushing us down the food chain towards a vegetarian diet if the limited area of inhabitable land that we have is to support the maximum number of people. So the size of human populations will be one of the factors determining how much you will pay for meat and how often it will appear on your plate.

Unlike the above simple example of a straight food chain, most consumers cross trophic levels. Consider ourselves. At how many trophic levels do we feed? Consider steak and salad. The steak comes from a herbivore and the salad is a producer. Suppose we eat a tuna sandwich. Tuna eat other carnivores. The complex feeding patterns in a community are represented by food webs. Energy may even be passed from the dead to the living. Vast numbers of decomposers make a living in this way.

Life after death

On the world-wide scale of size and abundance of living things, decomposers are microscopically small but astronomically large in numbers. They are almost all bacteria and fungi, which secrete digestive enzymes on the dead before absorbing the products of decay. Cellulose-rich bodies of dead plants, protein-rich dead animals, together with their nitrogen-laden waste products like urea, are the main courses of the meals of decomposers. While making a living in this way, they release carbon dioxide and water into the environment while they respire. In addition, they have evolved to become the world's experts in recycling. Nitrogen, carbon, phosphorus, sulphur, and a multitude of minerals are kept at constant levels as a result of decomposers. Without their help, the world would by a corpse-strewn wasteland, deficient in fertile soil.

The relationship between death, decay and nitrogen will serve to illustrate the essential role of decomposers. The Masai, a proud race living in Kenya and Tanzania, by tradition are warrior-herdspeople. They eat little except milk and blood, drained from their cattle which they prize above all other things. They are, therefore, almost totally carnivorous. The lion shares its habitat with the Masai and also relies on animal protein which it finds in herds of antelope. The Masai, lion, and all the world's carnivores depend on plants indirectly to obtain their nitrogen, the essential element of amino acids which are the building blocks of protein. Since nitrogen makes up nearly 80% of the atmosphere, we could be excused for thinking that there would be no problem in using it directly. The fact is, however, that very few organisms can use atmospheric nitrogen. Nitrogen is a really boring gas – it simply does not enter into the chemical reactions which occur in cells. Chemists call it inert because of its lack of reactivity. The only way that plants can be supplied with nitrogen is in the form of soluble nitrates, which enter roots from the soil. This is where death is necessary for life because death and decay are the main means of returning nitrogen to the soil. When an organism dies, bacteria decompose it. The putrid smelly materials that result from decay are nitrogen-rich compounds produced by rotting protein and urea in urine. Some bacteria live on the decomposition products and change them directly into soluble nitrogen compounds, and eventually to nitrates. Other bacteria feed on the nitrates and recycle nitrogen gas. Still other soil-dwelling bacteria are able to change atmospheric nitrogen into soluble compounds which are absorbed by plants.

Essentially the same cycle happens in aquatic environments. Therefore, on land and in water, the ends of the nitrogen cycle are joined by death.

Live now, pay later

All the domestic cats that exist are obliged to survive on whatever planet Earth can provide for them. There is little risk of them abusing our planet's resources, though, because their needs are so few: a warm fireplace, some food and a little social acceptance. It is to their advantage that their requirements are so few because there simply isn't anything else for them.

They would face disaster if they depleted or destroyed their available resources. The point is that our world has limited resources which have to be shared by all its inhabitants. However, the influence of all the cats in the world on the delicate webs of life is just a gentle touch compared to the sledgehammer blows of human impact.

Earth's resources come in two categories; renewable and non-renewable. Renewable resources are those that can reproduce themselves. If we cut down trees, theoretically we can replace them with seedlings which eventually will reproduce more trees. Non-renewable resources, as the name implies, exist in finite amounts. Once they are used, they are gone for ever. The exceptions to this rule are those that can have an extension to their existence by being recycled, but even these may be difficult to recover in their original form.

Food and water in focus

A consideration of renewable resources inevitably focuses on the provision of adequate food and water for the world's increasing population.

We are placing a massive demand on resources that we have inherited from our ancestors who lived in a more sparsely populated and a less technological kind of world. Our numbers are increasing, but our resources are decreasing – the problem is as simple as that. The solution to it is the most difficult puzzle that humans have ever had to solve. Of course, eventually, Mankind must accept inevitable extinction. All we can do is slow down the process by a few thousand years or so – the time it takes for a sneeze on the geological time-scale. Despite the fact that we are the most successful species in our solar system, success, just like glory, is fleeting in geological terms.

The second problem is the good fortune of some and the bad luck of others depending on where they were born and bred. This is because of the unequal distribution of the earth's wealth. The middle eastern countries have the bulk of the earth's oil, the South Africans have 55% of the world's supply of gold and most of the chromium and diamonds. America and Canada have most of the trees. This unequal distribution of desirable materials has been the cause of more wars than anything else except, perhaps, religious dogma. The opportunity to achieve world peace through mutual dependence of trade has been largely missed. Perhaps more attention should be given to the application of biology to solving the problem. Biological principles like competition, territoriality, inter-relationships and so on apply to our own species as well as others. Just like all species, we are the product of the blind and unprejudiced selection of genes of countless generations by pressures of the environment.

It may be hard for most people in the western developed world to understand the experience of starvation when their definition of hunger may be a late lunch after a missed breakfast. In fact, within recent times, Europeans and Americans have witnessed food mountains and situations where milk has been poured away due to overproduction. There is also an obsession with trivial detail regarding preference of certain foods and rejection of others with exactly the same nutritional value. There are critical standards for shapes, sizes and appearance for just about all fruits and vegetables which have to be met before they are saleable in European and American markets. Some argue the merits of red meats, fish and green vegetables. We often select just one part of a plant and dump the rest. Interestingly, many exponents of culinary fetishes are appallingly ignorant about nutrition; alarmingly often, if it is to the advantage of an entrepreneur that the fetishes become international. Many are told that 'natural' is always best and they end up believing it. Some actually believe that rose hips or blackcurrants, weight for weight, have more vitamin C than do chemically synthesised tablets of ascorbic acid. It is simply not so. Nor is organically grown food better for you, nutritionally, than food grown with chemical fertilizers. Slowly people are waking up to realise that food fetishes are expensive luxuries that increase the problems of world food supply. The food imbalance throughout the world is too great and the human population is growing too fast to allow us to continue 'business as usual'.

The problem of food shortage might be tackled if humans weren't so traditional and conservative in their tastes. It is very difficult to introduce a new food to a culture. No matter how nutritious the food is, people just won't eat it. Examples of this were seen in certain synthetic foods produced from fungi in biotechnology. Initial marketing problems were immense and in the case of mycoprotein, it took almost 30 years for the product to become moderately acceptable in the diets of people living in the UK.

There are over 50,000 species of known edible plants, but only 16 of these yield almost all the energy and protein that humans obtain from plants. In fact, most of this comes from the big three supergrasses; rice, wheat and corn. Amazingly, the rest of the edible species are ignored, even in the face of hunger and starvation! Incidentally, government nutritionists divide the hungry into two groups: the undernourished or those suffering from starvation, and the malnourished or those who do not have a balanced diet. Does the name of category really matter? In either case, lives are threatened.

A balanced diet is particularly important to children, pregnant women and nursing mothers. The women are essentially eating for two and so the risk of malnutrition is obvious. Children are particularly vulnerable to two very dangerous deficiency diseases, marasmus and kwashiorkor. Emaciation, coupled with bloated body cavities, wrinkled skin and a startling appearance of aging are symptoms. Both diseases could be prevented by extending nursing, but the mothers themselves require a good diet that may not be available.

The worldwide distribution of food is very patchy. Not only are there poor, hungry nations, and rich, well-fed ones, but the same country may have rich and poor parts. Southern Brazil is rich and fertile with affluent and well-fed citizens, while people in the less well off northern part of the country are more likely to be undernourished. The problem also exists in the USA. As late as the mid 1980s, it was estimated that 10–15% of Americans, mostly in impoverished inner-city ghettos and in western deserts, were malnourished, while others not that far away were choosing exotic sauces from over-priced gourmet menus.

The real world

The complexity of food problems again become apparent when we realised that on a global scale it is not just a matter of finding means to feed people. Food has its political implications. It is often used as a bargaining chip, as is the case for oil. Although the latter is perhaps a more powerful one and confrontations over its price have actually led to military intervention and at least one major war. The USA now sells wheat to Russia and other eastern European countries, but in the 1970s, because of political tension between these two great nations, the Americans refused to sell their surplus grain to the Soviet Union.

Scientists are looking towards the brave new world of genetic engineering as a branch of biotechnology to maximise food production from plants and animals that are already being cultivated or domesticated. Geneticists now have the ability to alter the genetic make up of plants and animals. They can make plants resistant to herbicides, insect pests and disease. They can improve the yield of almost all the organisms that we use as food. Undoubtedly the 'new biology' has the potential for increasing food production. Unfortunately, the public image of scientists and multi-national industrialists, making profits from genetically modified plants and animals, has been used by the media in a biased way to promote a 'fear of the unknown' syndrome. In the same way, the introduction of new synthesised food products or 'new' types of food plants or animals has found opposition when introduced to a culture. There is often reluctance to accept a change in diet. Indeed, sometimes in politically incorrect ways, we associate stereotypes of races by the food they eat, but to suddenly change attitudes to food after thousands of years of culture is asking too much of marketing skills. The nearest thing that the world has to a universally accepted 'food' is probably Coca Cola!

Historically, the hungriest people in the world are those that recognise the fewest types of food. Of course this is not their fault because most of them have never been introduced to more than their cultural staple diet. In fact, even people in developed countries are warned to stock familiar foods for emergencies because even such 'sophisticates' have been known to starve rather than try something new. We would all probably wait days before opening a tin of our least favourite food!

It is interesting that of the thousands of species with which we share this planet, so few have been considered worthy of eating. In fact, with all our abilities and technology, modern humans have discovered virtually no new crop plant or domesticable animal within the last thousand years. The last time it was tried on a multinational scale was when 'Breadfruit Bligh' of HMS Bounty was sent to the Pacific in the eighteenth century to collect a new crop plant, the breadfruit, and introduce it to the West Indies. It was a potentially cheap way of feeding the slaves! It never got much further that this though. How many of us have eaten breadfruit today? 'Miracle', high yielding grains and larger than normal trout and salmon are simply variations of species that have been with us for thousands of years.

Water and the coming crisis

Our use of water is immense. It is estimated that each person in the western world uses, on average, 1000 litres per day. On the other hand, a person in an underdeveloped country is likely to use ten times less than this. Much water is used for irrigation, power generation and the industrial production of just about all of the goods that surround you. As a result of this prodigious use, many ecologists predict severe shortages in this century. The problem is pressing many countries even now, including certain states in the most affluent regions of the world. The problem is aggravated because not only is water running out, but much of the remainder is becoming dangerously polluted and unusable. Localized water shortages have forced agriculturalists and land developers into diverting waterways from their natural courses to the more thirsty areas where irrigation is needed. The ecological consequences of this could be just adding more problems to society. Increasing populations mean larger cities and beneath these, many critical water supplies are drying up.

In addition, industrial dumping of various chemical effluents into waterways render vast volumes of water unusable. In some coastal areas, underground supplies of fresh water are steadily encroached upon by salt water as the fresh water is pumped out. Superimposed on all of this, we now have reservoirs that are being filled by acid rain – a legacy of our industrial wealth and progress.

The only practical long-term solution seems to be careful conservation of the remaining dwindling supplies. We may be forced to change our eating habits to foods that require less water. Perhaps we may be asked to give up meat because grain to feed the animals takes a great deal of water to grow. We may be asked to pay even more than we do now for water in order to make us aware of the fact that most of us are water wasters.

The more we use, the more we lose

The line between renewable and non-renewable resources is cobweb thin. For example, water is considered cyclic in that what goes up in evaporation must come down in precipitation. It is therefore renewable. However, much water is tied up in various manufactured and natural products and is not easily retrievable. We also know that water may become so polluted that it is no longer usable. Thus in a sense, it is at least partly non-renewable.

Non-renewable resources are not distributed evenly over the world and, consequently, their use may be manipulated by nationalistic policies. The 'good life' for many people is based quite simply on more and more goods and services. As they climb the long, rather shaky, ladder of success, they utilise disproportionate amounts of materials and energy. Such artificial demands place a great burden on our reservoir of natural resources and create great amounts of waste that must be disposed of by the expenditure of yet more materials and energy.

Energy, you will recall, is simply the capacity to do work. We need it to be able to rearrange our natural resources to build places to shield us from the harsh extremes of the environment and to make all the artifacts that are known to Mankind. What non-essential item of civilisation violates your sensibilities most? Do we really need to use fossil fuels to light all of the neon advertising signs in the world? Could we exist without electric toothbrushes and tin-openers? How many tonnes of packaging goes straight into the bin each day with the obligatory six round-headed pins for each new shirt? The list is endless. There are heavy costs to our environment, both in making energy available and in using it, and these costs must be carefully weighed in calculating the net benefit we receive.

The Stone Age did not end because people ran out of stones. It ended because we found something better. Will our 'oil age' end in the same way? Oil companies have invested heavily in developing alternatives like photovolteic solar cells, but while there is still oil to sell, they have little incentive to bring them to a world market.

The vast majority of the energy that we use comes from fossil fuels. They are the legacy of the fate of ancient plant life. The energy from sunlight, trapped by the chloroplasts of these primeval plants has waited patiently to be released for hundreds of millions of years. By switching on your car ignition or your gas pilot light, you will do the job in less than a second.

The earliest aquatic plant life has given us oil and gas by dying and accumulating 'en masse' in ancient seas and lakes. Their great weight impeded the natural process of decay and their partially decomposed biomass produced the liquified or gaseous fuel that has become the life-blood of the developed world. Our insatiable thirst is placing inordinate demands on these reserves even though they appear to be never-ending. It is interesting to think that the fate of the culmination of evolution may have been predetermined by lowly single-celled organisms in the palaeozoic seas of old.

Furthermore, in addition to being concerned about the availability of each energy source, we should consider the consequences of their exploitation. The effects of global warming and acid rain are already being felt. It has been estimated that billions of tonnes of airborne wastes reach the atmosphere each year. This is a staggering amount of pollution and has led nations to develop laws banning an increase in gases produced by industrial combustion.

Governments aim to reduce this form of pollution to a fraction of its present amount. We are aware of the problem when the TV weather forcaster gives us information about air quality. In some countries people are advised to stay indoors and remain inactive at times when there are very high levels of air pollution.

During four windless days in December 1991, air pollution in London reached a level that was the highest ever recorded in Britain. Scientists estimated that it might have been responsible for the deaths of up to 160 people because it aggravated lung problems, particularly in the elderly. The main pollutant was nitrogen dioxide which reached 423 parts per billion. This gas, together with other oxides of nitrogen and sulphur dioxide are all products of combustion of fossil fuels. The main sources of sulphur dioxide are coal-burning power stations. A major problem is that sulphur dioxide and oxides of nitrogen from traffic exhausts dissolve in rain to produce 'acid rain'. This has been the cause of devastation of forests and other plant life by causing toxic metal ions to separate from salts which occur naturally in soils. Acidification of lakes and streams also kill their inhabitants either directly or indirectly.

Stonework becomes corroded as acids dissolve it, causing millions of pounds worth of damage. Many famous classical sculptures from ancient civilisations have stood the test of time in the open air for thousands of years but are now kept inside, not only as protection against theft and vandalism, but also because of acid rain which is slowly corroding the smiles from the faces of the famous.

Acid rain has progressively thinned the shells of the eggs laid by wild birds over the past 150 years. Scientists fear that the trend could continue and make eggs less likely to hatch. The acid rain reduces the calcium content of the food of the birds so that the calcium available to form the egg shell is reduced.

All carbon-containing substances produce carbon dioxide when burned. Since the nineteenth century, it has been shown that there has been a massive increase in the amount of carbon dioxide in the atmosphere. This has been largely due to the increase in burning of fossil fuels and the destruction of much of the plant life of the world that would normally use the carbon dioxide for photosynthesis. As a consequence, there has been an accumulation of carbon dioxide as a layer around our planet. Radiation from sunlight passes through the atmosphere and heats up the Earth, but radiation reflected back by the Earth does not have enough energy to pass back through this layer. The four warmest years worldwide were all in the 1990s, with nine of the ten warmest occurring since 1980. It has been estimated that global warming by 4°C will occur in the next 20 years. The phenomenon is commonly known as the 'greenhouse effect' and, if predictions are accurate, will result in tremendous climatic changes and melting polar ice caps with disastrous consequences.

A popular misconception is that global warming is caused by damage to the ozone layer. Although ozone depletion is a very important problem, it is caused by a form of air pollution which is quite different to accumulation of carbon dioxide. The breaking down of ozone is caused by a build up of chlorofluorocarbons (CFCs) from aerosols and refrigerators. Ozone prevents cancer-inducing ultraviolet rays from the Sun from reaching harmful levels. Harmless alternatives to CFCs are now used wherever possible. In 1985, the world produced a million tonnes of ozone-destroying CFCs. By 1995, despite the fact that the use of CFCs was banned by industrialised nations, world production was still 360,000 tonnes. This was because developing nations are allowed to continue producing them until 2010.

CONCLUSION

Since modern scientific man is the only animal that has to adapt through will rather than through fate, it is important that he understands the rules by which to play the game. Thus, whether man shall remain the 'darling of the gods' or become extinct is strictly up to him.

H. H. Iltis (born–1925)

Each day we can, and do, make decisions that influence biological systems. You and I were once part of a future generation. Our ancestors worked to make the world a safer place for us. The future, however, soon becomes the present, and now we are here, we are the legacy of generations who never knew us. We are indeed gathering the harvest of the seeds that our ancestors sowed and there are both positive and negative aspects of our inheritance. In some ways, we benefit greatly from their concern and foresight. Testimony of this is in the higher standard of living that many of us enjoy. In other ways we have been left with an invoice for our forebears' expensive lifestyles in using non-replaceable resources. However, we should not be judgemental about the way they made their choices because there is no way that they could have imagined today's world.

It is now up to us to look to the future, but we too, are dazzled by the light of modern technology and cannot see ahead with any degree of certainty. Having used this as an excuse, we are to be forgiven less readily for our errors. We must bear greater responsibility than our ancestors because we know more than they did. We are aware of more potential problems on a global scale than any previous generation. We have enough information to understand the implications of our decisions. Of course we cannot predict what the future holds, but we can influence the future to an extent that could not have been imagined by our ancestors.

Undoubtedly, with this ability comes an added responsibility. We ignore our biological future at our peril. After 4,000 million years of evolution we are newly arrived delinquents with no insurance cover for the future. Unless we wish to foreclose the future of our descendants, we must make biological decisions that are conducive, not only to perpetuating life, but to improving its quality. The answer may partially lie in educating our law makers in the biological implications of their decisions and policies. Or perhaps the answer lies in first educating ourselves and then electing law makers from among those who are well aware of what they are doing and who can confidently expect the support of an enlightened public.

Rather than being accused of degrading the future of our descendants, let us hope that generations to come will be able to admire our objectivity and wisdom that we used when we made the decisions that pre-determined their fate.

A SHORT GUIDE TO BIOSPEAK

Many people find the technical vocabulary of biology off-putting. Indeed, some cannot see the point of speaking in ancient tongues when referring to anatomical terms or to names of organisms. The reason why biology is still liberally sprinkled with ancient words is because, several hundred years ago all science was taught through the medium of ancient classical languages. We are left with vestiges of this in an attempt at international communication. Scientific names of plants and animals are used universally so that scientists can communicate precisely the species used in their research. Anatomical terms were first used by the ancient Greeks and Romans as they were the first to name the parts that they observed.

The following list of the more important ancient Greek and Latin roots used in biological terminology has been compiled to enable the reader to understand the meaning of some of the most commonly used technical terms. They are arranged as prefixes and suffixes according to their usual use, though some may occur in either position. Their derivation from ancient Greek (G) or Latin (L) has been indicated.

Prefixes

a-	not, without (G)	anti-	against (G)
ab-	from (L)	arch-	primitive (G)
ad-	to, towards (L)	auto-	self (G)
amphi-	around (G)	aux-	growth (G)
ana-	up (G)	bi-	twice (L)
andro-	male (G)	bio-	life (G)
angio-	vessel (G)	botan-	herb (G)
ante-	before (L)	brachy-	short (G)

bryo-	moss (G)	hetero-	different (G)
carp-	fruit (G)	hexa-	six (G)
chaet-	bristle (G)	holo-	entire (G)
chloro-	green (G)	homo-	alike (G)
chondro-	grain-like (G)	hydro-	water (G)
chromo-	colour (G)	hyper-	above (G)
circum-	around (L)	hypo-	below (G)
co-, con-	together (L)	inter-	between (L)
coeno-	in common (G)	intra-	within (L)
coleo-	sheath (G)	iso-	equal (G)
crypto-	hidden (G)	karyo-	nut, nucleus (G)
cyano-	blue (G)	leuco-	white
cyto-	cell (G)	lys-	loosen, break (G)
derm-	skin (G)	macro-	large (G)
di-	twice (G)	mega-	large (G)
dia-	through (G)	meio-	less (G)
dicho-	separate (G)	melan-	black (G)
diplo-	double (G)	meris-	part (G)
e-	without (L)	meso-	middle (G)
ecto-	outside (G)	meta-	with (G)
endo-	inside (G)	micro-	small (G)
entomo-	insect (G)	mito-	thread (G)
epi-	upon (G)	mon-	single (G)
erythro-	red (G)	morph-	shape (G)
eu-	well, proper (G)	multi-	many (L)
ex-	out of (L)	myco-	fungus (G)
gam-	marriage (G)	necro-	dead (G)
gamet-	spouse (G)	neo-	new (G)
gastero-	stomach (G)	oligo-	few (G)
ge-	earth (G)	omni-	all (L)
geno-	racial (G)	oo-, ovo-	egg (G) (L)
gluc-, gly-	sweet (G)	ornitho-	bird (G)
gymno-	naked (G)	ortho-	straight (G)
gyn-	female (G)	pachy-	thick (G)
halo-	salt (G)	palaeo-	ancient (G)
haplo-	single (G)	pan-	all (G)
helio-	sun (G)	para-	beside (G)
hemi-	half (G)	penta-	five (G)

peri-	around (G)	scler-	hard (G)
pheno-	appearance (G)	semi-	half (L)
phloe-	tree bark (G)	soma-	body (G)
photo-	light (G)	sperma-	seed (G)
phylo-	tribe (G)	sub-	below (L)
pinna-	feather (G)	super-	above (L)
pleio-	more (G)	syn-	together (G)
poly-	many (G)	telo-	end (G)
post-	after (L)	thalasso-	salt water (G)
pre-	before (L)	therm-	heat (G)
pro-	for (L)	trans-	across (L)
proto-	first (G)	tri-	three (L) (G)
pseudo-	false (G)	tricho-	hair (G)
ptero-	wing (G)	uni-	one (L)
quadri-	four (L)	xero-	dry (G)
rhiz-	root (G)	xyl-	wood (G)
rhodo-	red (G)	zoo-	animal (G)
sapro-	putrid (G)	zygo-	joined (G)
schizo-	splitting (G)		

Suffixes

-androus	male (G)	-morphic	shaped (G)
-blast	bud, rudiment (G)	-phage	eating (G)
-carp	fruit (G)	-philous	loving (G)
-cyst	cavity (G)	-phobic	hating (G)
-cyte	cell (G)	-phyll	leaf (G)
-derm	skin (G)	-phyte	plant (G)
-enchyma	infusion (G)	-plasm	moulded (G)
-ferous	bearing (L)	-rhiza	root (G)
-fid	cleft (L)	-scopic	looking (G)
-folium	leaf (L)	-sperm	seed (G)
-gen	producing (G)	-tactic	arranged (G)
-generous	bearing (L)	-trophic	nourished (G)
-gynous	female (G)	-tropic	turned (G)
-kinesis	movement (G)	-vorous	consuming (L)
-logy	knowledge (G)	-zoid	animal-like (G)
-lytic	dissolving (G)		

GLOSSARY

Abdomen   The area between the thorax and the pelvis.

Actin   Slender filaments of protein arranged in bundles in the composition of muscle fibrils.

Active transport   The passage of a substance through a cell membrane requiring the use of energy.

Adaptation   The process in which a species becomes better suited to survive in an environment.

Adrenal glands   Two ductless glands located above each kidney.

Aerobic   Requiring free atmospheric oxygen for normal activity.

Aerosols   Tiny suspended droplets in air.

Air sacs   Thin-walled divisions of the lungs.

Alimentary canal   Those organs that compose the food tube.

Allele   One of a pair of genes responsible for contrasting characters.

Alveoli   Microscopic sacs in the lungs in which exchange of gases takes place.

Amino acids   Substances from which organisms build protein; the end products of protein digestion.

Amnion   The innermost foetal membrane, forming a sac surrounding the foetus.

Amniotic fluid   Secreted by the amnion and filling the cavity in which the embryo lies.

Anaerobic   Deriving energy from chemical changes other than involving oxygen.

Anther   That part of the stamen which bears pollen grains.

Antibiotic   A bacteria-killing susbtance produced by a microbe.

Antibody   A substance in the blood that helps lead to immunity.

Antigen   A substance, usually a protein, which when introduced into the body stimulates the production of antibodies.

Antitoxin   A substance in the blood that counteracts a specific toxin.

Anus   The opening at the posterior end of the alimentary canal.

Aorta   The largest artery in the body which leads from the heart.

Aquatic   Living in water.

Aqueous humor   The watery fluid filling the cavity between the cornea and the lens of the eye.

Arteriole   A tiny artery that eventually branches to become capillaries.

Artery   A large muscular blood vessel that carries blood away from the heart.

Asexual reproduction   Reproduction without the fusion of sex cells.

ATP (adenosine triphosphate)   A high-energy compound found in cells that functions in energy storage and transfer.

Atrioventricular valves   The heart valves located between the atria and the ventricles.

Atrium   A thin-walled upper chamber of the heart that receives blood from veins.

Auditory nerve   The nerve leading from the inner ear to the brain.

Autonomic nervous system   A division of the nervous system that regulates involuntary actions in internal organs.

Autosome   Any paired chromosome other than the sex chromosomes.

Autotroph   Organisms capable of organising inorganic molecules into organic molecules.

Auxin   A plant hormone that regulates growth.

Axillary buds   Buds formed in the angle between a leaf stalk and a stem.

Axon   A nerve process that carries an impulse away from the nerve body.

Bacteria   A group of microscopic organisms without nuclear membranes.

Bile   A brownish-green emulsifying fluid secreted by the liver and stored in the gall bladder.

Binary fission   The division of a single-celled organism into two similar cells.

Biodegradable   Material decomposed by natural processes.

Biomass   The mass of living material per unit area or volume.

Biome   A large geographical region identified mainly by its major vegetation.

Biosphere   The area in which life is possible on Earth.

Blastula   An early stage in the development of an embryo, in which cells have divided to produce a hollow sphere.

Bowman's capsule   The cup-shaped structure forming one end of the tubule and surrounding a knot of blood capillaries in the tubule of a kidney.

Brain stem   An enlargement at the base of the brain where it connects to the spinal cord.

Breathing   The mechanism of getting air in and out of the lungs.

Bronchiole   One of the numerous subdivisions of the bronchi within a lung.

Bronchus   A division of the lower end of the windpipe leading to a lung.

Bud   An undeveloped shoot of a plant, often covered by scales; developing young, forming on some simple animals such as *Hydra.*

Bulb   A form of underground stem composed of thick scale leaves; e.g. onion.

Calorie   A unit used to measure the energy in food. It is now outdated and should be substituted by the joule as a unit. A kilocalorie is the amount of heat needed to raise the temperature of 1 kilogram of water (1 litre) to one degree Celsius.

Cambium   A ring of dividing cells in roots and stems that forms new xylem and phloem.

Canine   Teeth for tearing.

Capillary   The smallest blood vessels in the body through which exchanges occur between blood and tissue fluid.

Cardiac muscle   Muscle composing the heart wall.

Carnivore   A meat eater.

Cartilage   A strong, pliable, smooth tissue which supports structures and lines bones at joints.

Catalyst   A substance that accelerates a chemical reaction without being altered chemically.

Cell   A unit of structure and function of an organism.

Cell theory   The cell is the unit of structure and function of all living things and arises from a pre-existing cell by division.

Cementum   The covering of the root of a tooth.

Central nervous system   The brain and spinal cord, and nerves arising from them.

Cerebellum   The brain region between the cerebrum and medulla, concerned with balance and muscular co-ordination.

Cerebrum   The largest region of the brain, considered to be the seat of emotions, intelligence, and voluntary activities.

Cervix   The neck of the uterus.

Chlorophyll   Green pigments essential to food manufacture in plants.

Chloroplast   A structure in a cell that contains chlorophyll.

Choroid layer   The layer of the eye beneath the sclera which prevents internal reflection and contains most blood vessels.

Chromatid   During cell division, each part of a double-stranded chromosome.

Chromosome   A rod shaped gene-bearing body in the cell, composed of DNA joined to protein molecules.

Cilia   Tiny hair-like projections of cytoplasm.

Ciliary muscles   Those that control the shape of the lens in the eye.

Cleavage   The rapid series of divisions that a fertilised egg undergoes.

Cochlea   The hearing apparatus of the inner ear.

Coenzyme   A molecule that works with an enzyme in catalysing a reaction.

Cohesion   The clinging together of molecules as in a column of liquid in transpiration.

Coleoptile   A protective sheath encasing the primary leaf of the oat plant and other grasses.

Colloid   A gelatinous substance, such as protoplasm or egg albumen, in which solids are dispersed throughout a liquid.

Colon   Part of the large intestine.

Cone   Cells in the retina responsible for colour-vision.

Connective tissue   A type of tissue that lies between groups of nerve, gland and muscle cells.

Consumer   An organism that feeds on another organism.

Contractile vacuole   A cavity in some single-celled organisms associated with the elimination of excess water.

Cornea   A transparent bulge of the sclera of the eye in front of the iris, through which light passes.

Corpus luteum   Refers to the follicle in an ovary after an ovum is discharged.

Cortex   In roots and stems, a storage tissue; in organs such as the kidney and brain, the outer region.

Cotyledon   A seed leaf present in the embryo plant that serves as a food store.

Cowper's gland   Located near the upper end of the male urethra. It secretes a fluid which is added to the sperms.

Cytoplasm   The protoplasmic materials in a cell lying outside the nucleus and inside the cell membrane.

Daughter cells   Newly formed cells resulting from the division of a previously existing cell, called a mother cell. The two daughter cells receive identical nuclear materials.

Decomposers   Organisms that break down the tissues and excretory material of other organisms into simpler substances through the process of decay.

Deficiency disease   A condition resulting from the lack of one or more vitamins.

Denitrification   The process carried out by denitrifying bacteria in breaking down ammonia, nitrites and nitrates, and liberating nitrogen gas.

Dentine   A substance that is relatively softer than enamel, forming the bulk of a tooth.

Depressant   A drug having an anaesthetic effect on the nervous system.

Diaphragm   A muscular partition separating the thorax from the abdomen. Also, a contraceptive device used as a barrier at the cervix.

Diastole   Part of the cycle of the heart during which the ventricles relax and receive blood from the atria.

Diffusion   The spreading out of molecules in a given space from a region of greater concentration to one of lesser concentration.

Digestion   The process during which foods are chemically simplified and made soluble for absorption.

Diploid   Term used to indicate a cell or an organism that contains a full set of homologous pairs of chromosomes.

DNA (deoxyribonucleic acid)   A giant molecule in the shape of a double helix, consisting of alternating units of nucleotides; composed of deoxyribose sugar, phosphates and nitrogen bases.

Dominance   The principle first observed by Mendel, that one gene may prevent the expression of an allele.

Duodenum   The region of the small intestine immediately following the stomach.

Ecology   The study of the relationships of living things to their surroundings.

Ecosystem   A unit of the biosphere in which living and non-living things interact, and in which materials are re-cycled.

Egestion   Elimination of insoluble, non-digested waste.

Egg   A female reproductive cell.

Embryo   An early stage in a developing organism.

Enamel   The hard covering of the crown of a tooth.

Endocrine gland   A ductless gland that secretes hormones directly into the bloodstream.

Endoskeleton   Internal framework of vertebrates made of bone and/or cartilage.

Environment   The surroundings of an organism; all external forces that influence the expression of an organism's genes.

Enzyme   A protein that acts as a catalyst.

Epiglottis   A cartilaginous flap at the upper end of the trachea.

Epithelium   A tissue that covers various organs and the body surface.

Etiolated   A condition of pale yellow leaves and stems of plants which are grown in the dark. The stems are also elongated and weak.

Eustachian tube   A tube connecting the throat with the middle ear.

Evolution   The slow process of change by which organisms have acquired their distinguishing characteristics.

Excretion   The process by which metabolic wastes are removed from cells and the body.

Exoskeleton   The hard outer covering of certain animals.

Expiration   The discharge of air from the lungs.

Extensor   A muscle that straightens a joint.

Faeces   Intestinal solid waste material.

Fallopian tube   See oviduct.

Fallout   Radioactive particles that settle on the Earth from the atmosphere.

Fermentation   Glucose oxidation that is anaerobic and in which lactic acid is formed in muscle and ethanol by plants.

Fertilization   The union of two gametes.

Foetus   Mammalian embryo after the main body features have formed.

Fibrin   A substance formed during blood clotting by the reaction between thrombin and fibrinogen.

Fibrinogen A blood protein present in plasma involved in clotting.

Flexor A muscle that bends a joint.

Food Any substance absorbed into the cells of the body that yields material for energy, growth and repair of tissue and regulation of the life processes, without harming the organism.

Food chain The transfer of the sun's energy from producers to consumers as organisms feed on one another.

Food pyramid A quantitative representation of a food chain, with the food producers forming the base and the top carnivore at the apex.

Food web Complex food chains existing within an ecosystem.

Fossil The imprint or preserved remains of an organism that once lived.

Fovea The most light-sensitive spot on the retina of the eye where cones are most abundant.

Fungus A plant-like organism that lacks chlorophyll and therefore derives nourishment from an organic source.

Gall bladder A sac in which bile from the liver is stored and concentrated.

Gamete A male or female reproductive cell.

Gastric fluid Glandular secretions of the stomach.

Gene That portion of a DNA molecule that is capable of replication and mutation and passes on a characteristic from parent to offspring.

Genetic code That sequential arrangement of bases in the DNA molecule, which controls characteristics of an organism.

Genetics The science of heredity.

Genotype The hereditary make up of an organism.

Geotropism The response of plants to gravity.

Germination The growth of a seed when favourable conditions occur.

Gill An organ modified for absorbing dissolved oxygen from the water.

Glomerulus The knot of capillaries within a Bowman's capsule of a kidney tubule.

Gonads The male and female reproductive organs in which gametes and hormones are produced.

Guard cell One of two epidermal cells surrounding a stoma on a plant.

Guttation   The forcing of water from the leaves of plants, usually when the stomata are closed.

Habitat   A place where an organism lives naturally.

Haemoglobin   An iron-containing protein, giving red blood cells their red colour and which carries oxygen around the body.

Haploid   A term used to indicate a cell, such as a gamete, that contains only one chromosome of each homologous pair.

Hepatic portal vein   A vessel carrying blood to the liver before the blood returns to the heart.

Herbivore   A plant-eating animal.

Heredity   The transmission of characteristics from parents to offspring.

Heterotroph   An organism that is unable to make organic molecules from inorganic molecules. It is dependent on other organisms for food.

Heterozygous   Refers to an organism in which the paired genes for a particular characteristic are different.

Homeostasis   A steady state that an organism maintains by self-regulating adjustments.

Homologous chromosomes   A pair of chromosomes which are identical in form and in the way in which genes are arranged.

Homozygous   Refers to an organism in which the paired genes for a particular characteristic are identical.

Hormone   The chemical secretion of a ductless gland producing a physiological effect.

Host   In a parasitic relationship, the organism from which the parasite derives its food.

Hydrotropism   The response of roots to water.

Hypothalamus   Part of the brain that controls the pituitary gland.

Hypothesis   A working explanation or trial answer to a question.

Ileum   The longest region of the small intestine where digestion is completed.

Immunity   The ability of the body to resist disease by natural or artificial means.

Incisor   One of the cutting teeth in the front of both jaws in mammals.

Insecticide   A chemical that kills insects.

Inspiration   The intake of air into the lungs.

Interferon   A cellular chemical defence against a virus.

Invertebrate   An animal without a backbone.

Involuntary muscle   One that cannot be controlled at will, like smooth muscle.

Iris   The muscular, coloured part of the eye behind the cornea, surrounding the pupil.

Islets of Langerhans   Groups of cells in the pancreas that secrete insulin.

Joint   The place at which two bones meet.

Kidney   An excretory organ that filters urea from the blood.

Lacteal   A lymph vessel that absorbs the products of fat digestion from the intestinal wall.

Larynx   The voice box.

Ligament   A tough strand of connective tissue that holds bones together at a joint.

Limiting factor   Any factor that is essential to organisms and for which an increase will cause an increase in the rate of a reaction.

Liver   The largest gland in the body, associated with metabolism of carbohydrate, protein and fat.

Lung   An organ for gaseous exchange during breathing.

Lymph   The clear liquid part of the blood that enters tissue spaces and lymph vessels.

Mammary glands   Those glands found in female mammals that secrete milk.

Medulla   In the kidney, the inner portion composed of pyramids, that in turn, are composed of tubules; in the adrenal gland, the inner portion.

Medulla oblongata   The enlargement at the base of the brain called the brain stem. It controls the activities of internal organs.

Meiosis   The type of cell division in which, in the production of eggs and sperm, there is a reduction of chromosomes to the haploid number.

Meninges   Protective membranes surrounding the spinal cord and brain.

Menstruation   The periodic breakdown and discharge of the uterine lining that occurs after puberty in the absence of fertilization.

Messenger RNA   The type of RNA that receives the code for a specific protein from DNA in the nucleus and acts as a template for protein synthesis on the ribosome.

Metabolism   The sum of the chemical processes taking place in the body.

Mitosis   The division of chromosomes preceding the division of cytoplasm and leading to two identical cells from an original cell.

Molar   A large tooth only present in an adult's dentition and used for grinding; sometimes called a wisdom tooth.

Motor end plate   The end of an axon of a motor nerve in a muscle.

Mucus   A slimy, lubricating and cleaning secretion from mucous glands.

Mutation   A change in genetic make up resulting in a new characteristic that can be inherited.

Myosin   A form of thick filamentous protein that, together with actin filaments, composes a muscle fibril.

Natural selection   The result of survival of the fittest to breed in the struggle for existence among organisms possessing those charcteristics that give them an advantage.

Nephron   One of the numerous excretory tubules in the kidney, including the Bowman's capsule and glomerulus.

Nerve cord   Part of the central nervous system extending from the brain along the dorsal side of the body.

Nerve impulse   An electrochemical stimulus causing change in a nerve fibre.

Neuron   A nerve cell body and its processes.

Nitrification   The action of a group of soil bacteria on ammonium compounds, producing nitrates.

Nitrogen cycle   A series of chemical reactions in which nitrogen compounds change form resulting in stable concentration of nitrogen in the atmosphere and nitrate in the soil.

Nitrogen fixation   The process by which certain bacteria in the soil or in the roots of leguminous plants change free nitrogen into nitrogen compounds that plants can use.

Non biodegradable   Polluting materials that are not decomposed by natural processes.

Nucleotide   A unit composed of deoxyribose sugar, a posphate, and a base. Many such units make up DNA.

Nucleus   The part of the cell that contains chromosomes and which controls its activities.

Optic nerve   The nerve leading from the retina of the eye to the optic lobe of the brain.

Organ   Different tissues grouped together to perform a function or functions as a unit.

Organism   A complete and entire living thing.

Osmosis   The diffusion of water through a selectively permeable membrane from a region of greater concentration of water to a region of lesser concentration of water.

Ovary   The female reproductive organ.

Oviduct   One of a pair of tubes in a female through which eggs travel from the ovary and in which fertilization occurs.

Ovule   A structure in the ovary of a flower that can become a seed when the egg is fertilised.

Palisade mesophyll   Leaf tissue composed of elongated cells in upright rows.

Pancreas   A gland located near the stomach and duodenum that produces digestive juices and hormones.

Parasite   An organism that lives in or on the body of another living organism called a host and obtains nourishment from it.

Pasteurization   The process of killing and/or retarding the growth of bacteria in milk and alcoholic drinks by heating to a selected temperature so that the flavour is retained.

Pathogenic   Disease-causing.

Pectoral girdle   The framework of bones by which the forelimbs of vertebrates are supported.

Pelvic girdle   The framework of bones by which the hind limbs of vertebrates are supported.

Pericardium   The membrane surrounding the heart.

Periodontal membrane   The fibrous structure that anchors the root of a tooth in the jaw socket.

Peripheral nervous system   The nerves communicating with the central nervous system and other parts of the body.

Petal   One of the coloured parts of the flower which attracts insects for insect pollination.

Phenotype   The outward expression of genes.

Phloem   The tissue in leaves, stems and roots that carries dissolved food substances.

Photosynthesis   The process by which plant cells combine carbon dioxide and water in the presence of chlorophyll and light, to form carbohydrates and release oxygen.

Pituitary gland   A ductless gland composed of two lobes, located beneath the cerebrum.

Placenta   A large, thin membrane in the uterus that exchanges materials between the mother and the embryo by diffusion.

Plankton   Floating organisms.

Plasma   The liquid part of blood.

Plasmolysis   Osmosis applied to single cells.

Platelet   The smallest of the solid components of blood, which releases thrombokinase for clotting.

Pollen   The structures containing the male sex nuclei produced in anthers of flowers.

Pollen tube   The tube formed by a pollen grain which carries the male nuclei to the female egg cell for fertilisation.

Pollination   The transfer of pollen from the anther to the stigma.

Pollution   The addition of impurities.

Population   A group of inter-breeding individuals in an ecosystem.

Predator   Any animal that preys on another.

Prostate gland   A gland located near the upper end of the urethra in a male, helping to produce seminal fluid.

Protoplasm   Refers to the complex, constantly changing system of chemicals in a cell that establishes the living condition.

Puberty   The age at which the secondary sex characteristics appear.

Pulmonary   Pertaining to the lungs.

Pulse   Regular expansion of the artery walls caused by the beating of the heart.

Pupil   The opening in the front of the eyeball, the size of which is controlled by the iris.

Pyloric sphincter   A ring of muscle regulating the passage of semi-liquids from the stomach to the duodenum.

Quarantine   Isolation of organisms to prevent the spread of infection.

Radioactive   Refers to an element that spontaneously gives off radiation.

Receptor   A cell or group of cells, that receives a stimulus.

Recessive   Refers to a gene or character that is masked when a dominant allele is present.

Rectum   The posterior portion of the large intestine.

Reduction division   See meiosis.

Reflex action   A nervous reaction in which a stimulus causes the passage of a sensory nerve impulse to the spinal cord or brain, from which, a motor impulse is transmitted to a muscle or a gland.

Renal   Relating to the kidney.

Replication   Self duplication, or the process whereby a DNA molecule makes an exact duplicate of itself.

Reproduction   The process through which organisms produce offspring.

Respiration   The release of energy from glucose in every living cell.

Response   The reaction to a stimulus.

Retina   The inner layer of the eyeball, formed from the expanded end of the optic nerve.

RNA (Ribonucleic acid)   A nucleic acid in which the sugar is ribose. A product of DNA, it controls protein synthesis.

Rod   A type of cell in the retina of the eye that responds to shades of light and dark but not to colours.

Root cap   A tissue at the tip of a root protecting the growing region.

Root hair   A projection of the outermost layer of a root which increases the surface area for absorption of water and minerals.

Root pressure   The pressure that is built up in roots due to water intake.

Saliva   A fluid secreted into the mouth by the salivary glands containing the enzyme, amylase.

Salivary gland   A group of secretory cells producing saliva.

Saprophyte   A fungus or bacterium that lives on dead organic matter.

Sclerotic layer (Sclera)   The outer layer of the wall of the eyeball.

Scrotum   The pouch outside the abdomen that contains the testes.

Sedative   An agent that depresses body activities.

Seed   A complete embryo plant surrounded by a food store and a protective coat.

Selectively permeable membrane   One that lets some substances pass through more readily than others depending on their molecular size.

Semen   Fertilizing fluid consisting of sperms and fluids from the seminal vesicle, prostate gland, and Cowper's gland.

Semicircular canals   The three curved passages in the inner ear that are associated with balance.

Semilunar valves   Cup shaped valves at the base of the aorta and the pulmonary artery that prevent back flow of blood into the ventricles.

Seminal vesicles   Structures that store sperm.

Seminiferous tubules   A mass of coiled tubes in which sperms are formed in the testes.

Sensory neurons   Those that carry impulses from a receptor to the central nervous system.

Sepal   The outermost part of a flower, usually green and protecting the bud.

Serum   Plasma without clotting factors.

Sex chromosomes   The two kinds of chromosomes (X and Y) that determine the gender of a person.

Sex linked character   A recessive character carried on the X chromosome.

Small intestine   The digestive tube, about seven metres long that begins with the duodenum and ends at the colon.

Smog   Combination of smoke and fog.

Smooth muscle   That which is involuntary and is found in the walls of the intestine, stomach and arteries.

Sperm   Short for spermatozoon – a male reproductive cell.

Sphincter muscle   A ring of smooth muscle that closes a tube.

Spinal cord   The main dorsal nerve of the central nervous system.

Spinal nerves   Large nerves connecting the spinal cord with the organs of the body.

Spongy mesophyll   Loosely connected leaf tissue with many air spaces.

Stamen   The male reproductive organ of a flowering plant.

Stigma   The part of the female reproductive organ of a flowering plant that receives pollen during pollination.

Stimulant   An agent that increases body activity.

Stoma   A pore regulating the passage of air and water vapour to and from leaves and stems.

Stomach   An organ that receives ingested food, prepares it for digestion, and begins protein digestion.

Style   The stalk of the female reproductive organ of a flowering plant.

Symbiosis   The relationship between two organisms of different species for mutual benefit.

Synapse   The space between nerve endings.

Synovial fluid   A secretion that lubricates joints.

Systole   Part of the cycle of the heart during which ventricles contract and force blood into arteries.

Taste buds   Flask-shaped structures in the tongue containing nerve endings that are stimulated by chemicals.

Tendon   A strong band of connective tissue which connects a muscle to a bone.

Testa   The outer seed coat.

Testes   The male reproductive organs which produce sperms.

Thorax   The middle region of the body between the neck and the abdomen.

Thrombin   A substance formed in blood clotting as a result of the reaction of prothrombin, thrombokinase, and calcium ions.

Thrombokinase   A substance essential to blood clotting formed by breakdown of platelets.

Thyroid   The ductless gland, located in the neck, that regulates metabolism.

Tissue fluid   That which bathes cells of the body. It is called lymph when in vessels.

Trachea   The windpipe taking air to the bronchi.

Transfer RNA   A form of RNA which delivers amino acids to the template formed by messenger RNA on the ribosomes.

Translocation   The movement of manufactured, dissolved food substances in the phloem.

Transpiration   The loss of water from plants.

Transpiration pull   The pressure which forces water to rise in a stem, as cells lose water to the atmosphere by transpiration.

Tropism   A growth response in a plant due to a uni-directional stimulus.

Turgor   The stiffness of plant cells due to the pressure exerted by contained water.

Tympanic membrane   The ear drum.

Umbilical cord   The link between the embryo and the placenta.

Urea   A nitrogenous waste substance made in the liver from excess amino acids.

Ureter   A tube leading from the kidney to the bladder.

Urethra   A tube leading from the bladder to an external opening of the body.

Urine   The liquid waste made in the kidney and stored in the bladder, consisting mainly of water and urea.

Uterus   The organ in which developing embryos are nourished and protected until birth.

Vaccination   Method of producing immunity by inoculating with a vaccine.

Vaccine   A substance used to produce immunity.

Vacuole   One of the spaces scattered through the cytoplasm of a cell and containing liquid.

Vagina   Cavity in the female immediately outside and surrounding the cervix of the uterus.

Vas deferens   Tubes that transport sperms from the testes.

Vascular bundles   Strands of phloem and xylem found in roots, stems and leaves for transporting materials in solution around the plant.

Vector   A carrier of disease-causing organisms.

Ventricle   A muscular chamber of the heart for pumping blood.

Vertebra   A backbone.

Vertebrate   An animal with a backbone.

Villi   Microscopic projections of the wall of the small intestine or the placenta for increasing the surface area.

Viruses   Particles that are non-cellular and have no nucleus or no cytoplasm. They cannot reproduce unless they are inside living cells.

Vitamin   An organic substance tht helps enzymes work in the body.

Vitreous humor   A transparent jelly-like substance that fills the interior of the eye ball and maintains its shape.

Vocal cords   Those structures within the larynx that vibrate to produce sound.

Voluntary muscle   Striated muscle that can be controlled at will.

White blood cells   Colourless nucleated blood cells for defence.

X chromosome   A sex chromosome present singly in males and as a pair in females.

Xylem   The woody tissue of plants that conducts water and dissolved minerals.

Y chromosome   A sex chromosome found only in males.

Zygote   The product of fertilization.

INDEX

TEACH YOURSELF

GENETICS

Morton Jenkins

Teach Yourself Genetics provides an essential introduction to this ever-changing area of science. Whatever your particular interest, from the basic science of genetics and inheritance to the ethics of genetic engineering, eugenics and the Human Genome project, this is the book for you.

- Take a closer look at Mendel's pioneering work.
- Examine the applications of basic concepts to modern gene technology.
- Uncover modern genetic developments.
- Explore the historical theory of genetics.

Morton Jenkins is an experienced teacher and author in the biological field.

TEACH YOURSELF

HUMAN ANATOMY & PHYSIOLOGY
4th Edition

David Le Vay

This practical and comprehensive guide provides an ideal introduction to the structure and function of the human body. The composition and properties of body tissues are fully described together with the anatomy of the various body systems and their component tissues. The second part of the book focuses on physiology and related biochemical and biophysical processes.

Extensively illustrated, the book covers:

- modern methods of investigation of structure and function
- relevant aspects of modern genetics
- aspects relating to physical training and sports injuries
- environmental and evolutionary considerations
- physiological aspects of AIDS.

David Le Vay MS, FRCS was a consultant surgeon for many years and is now a well known medical author and editor.

Other related titles

CONSERVATION

Nicholas Foskett & Rosalind Foskett

Conservation is one of the key issues in science today, affecting both the world around us and our lifestyles. *Teach Yourself Conservation* provides an essential introduction to this fascinating subject. The history, nature and issues of conservation are all explored and explained, with emphasis on the implications of conservation for the future.

The book:

- covers the environment and animal kingdom
- gives up-to-date and contemporary coverage of conservation issues
- covers issues on a local and global scale
- shows the possible ways forward for world conservation.

Nicholas Foskett and Rosalind Foskett are both lecturers in Environmental Science and Education. They have published and researched extensively in these fields.

TEACH YOURSELF

GEOLOGY

David A. Rothery

We all live on the Earth, but how many of us are aware of all the processes that shape – and sometimes shake – its surface? *Teach Yourself Geology* is a comprehensive introduction to the nature and history of the Earth, ranging from volcanoes to fossils, from the problems of living with earthquakes to the implications of limited natural resources.

Featuring extensive illustrations, the book covers:

- the origin and evolution of the Earth
- rocks, minerals and fossils
- key geological processes
- earthquakes and volcanoes
- geology on other planets
- how to carry out fieldwork.

David Rothery's research work has taken him to places as diverse as Western Australia, the Oman mountains of Arabia, and the volcanoes of South America and Hawaii. He has also worked on the geographical interpretation of spaceprobe pictures of Mars and the satellites of outer planets.

TEACH YOURSELF

VOLCANOES

David A. Rothery

Volcanoes are spectacular and can have devastating effects. As well as the obvious death and destruction surrounding an erupting volcano, the largest eruptions can change the world's climate. This practical comprehensive guide will enable you to discover more about the mysteries behind volcanic activity.

Extensively illustrated, the book:

- explains why volcanoes occur and why they erupt
- describes the various kinds of volcanoes on the Earth and on other planetary bodies
- discusses the hazards posed by volcanoes, both locally and globally
- explains why volcanoes can vary in shape
- shows how volcanic activity can be monitored and predicted.

David Rothery is a volcanologist at the Open University. His international collaborative research on active volcanoes has taken him to places as diverse as Hawaii, the Americas, Indonesia and Italy.